Breeding Better Bees

Using Simple Modern Methods

BIBBA is an educational charity founded in 1964 by the late Beowulf Cooper. Originally called the Village Bee Breeders' Association, its aims remain unchanged as:

'The conservation, restoration, study, selection and improvement of our native and near native honeybees of Britain and Ireland.'

BIBBA seeks to further these aims by the production and sale of relevant publications, by holding conferences and other practical bee breeding meetings and, above all, by encouraging members to work together in local groups on selection and breeding schemes for the improvement of their own bees.

Breeding Better Bees

Using Simple Modern Methods

John E Dews and Eric Milner

3rd edition, 2004

Printed and bound in the UK by PublishPoint
from KnowledgePoint Limited, Reading

© *John E Dews, Eric Milner 1993*

This book is sold subject to the condition that it shall not, by way of trade or otherwise, be lent, resold, hired out, converted to another format or otherwise circulated without the publisher's prior consent in any form of binding or cover other than that in which it is published. The moral right of the author has been asserted

First published in Great Britain 1993
by British Isles Bee Breeders Association

Third edition 2004 *with WritersPrintshop*
ISBN 1904623182

William Henry Paul Gould

When Paul Gould died in 1990, BIBBA lost a valued member and friend. He was a Life Member of the Association and a regular attender at conferences.

He was always generous to BIBBA, often making contributions to help the work of the Association. This extended to his will in which he left BIBBA a generous bequest. This bequest is being used for the publication of this booklet.

He always strived towards improving his bees and his quiet unassuming manner was a feature of his character – a beekeeper who will be missed by his friends.

It is with great pleasure that we dedicate this booklet to him.

Contents

1. **Introduction**	1
2. **The causes and effects of mongrelisation**	3
3. **Social problems caused by mongrel bees**	5
Bad temper	5
Swarms	5
4. **Economic and management problems**	7
5. **The German experience**	8
Productivity of bees in Germany	8
Pure race breeding	9
Danger of relapse	10
Racial hybrids	10
6. **The best bee for the British climate**	11
7. **The survival of the Dark bee in the British Isles**	15
8. **Principles of selection**	18
9. **Morphometric identification of the European races of honeybee**	22
Methods	23
Table for comparison of the main external characters of the four principal races of European honeybees	24
Taking of samples	25
Measurement of body characters	25
The wing measurements	26
The Cubital Index (CI)	26
The Discoidal Shift (DS)	29
Overhairs	30
Body colour	33
Tomenta	34
Proboscis	35
Interpreting the findings	36
Scattergram of Cubital Index and Discoidal Shift	37
10. **Queen cell production in a single brood chamber colony**	40
The timetable	42
11. **Making up and use of three-frame nuclei for queen mating and introduction**	43
Causes of failure of three-frame mating nuclei	43
Queen introduction using a three-frame nucleus	44

12. The organisation of a local breeding programme	46
Displacement breeding	47
Some ways of achieving pure matings of native bees	48
Distribution of breeding material	49
A brief note on the management of the native bee	50
13. Varroa and the bee breeder	52
14. Measuring the Cubital Index and Discoidal Shift with the aid of a computer program	55
References	63
Acknowledgements	64

Chapter 1
Introduction

Since 1859 many thousands of queen honeybees of foreign race have been imported into this country with the apparent intention of producing better bees than those already here, but it can be shown that this has had the opposite result.

It is our considered opinion that the major problem of beekeepers at the present time is the excessive number of mongrel bees in the country. Many of these mongrels are bad tempered, swarmy, relatively unproductive, unthrifty, and unsuited to cope with the vagaries of the climate. This poses social problems for the beekeeper and his neighbours and economic and management problems for the beekeepers. Unfortunately neither of these problems is as yet recognised by many in the beekeeping community. The prime cause of the problem is the continued importation of foreign bees.

There is a feasible solution. We need only to look to our German neighbours. Up to the end of World War II, the same situation existed there: the bees were 'sting happy', swarmy, unproductive mongrels. Now their bees are very gentle, disinclined to swarm, and much more productive. Official figures show that more honey is produced each year in former West Germany than in the whole of Germany before the war, in spite of the fact that fewer bees are now kept in former West Germany, and the forage conditions are worse, a deterioration largely caused by changes in agricultural practice.

All this has been achieved by Pure Race Breeding of a bee that is suited to the climate and does what the beekeepers require.

Professor Ruttner states that, if their present breeding methods were to be abandoned, the bees in Germany would sink back into the previous state of chaos from which they have been delivered.

In this country the native honeybee, *Apis mellifera mellifera*, ie, the Dark European honeybee, is the bee most suited to the climate. Contrary to popular belief, it still exists in a pure state and can be distinguished from mongrel bees and other races. With a **National Bee Breeding Programme** based on the German system, these bees, some of which already have the good qualities most beekeepers require, could be selectively bred and improved with an expectation of the same degree of success that the Germans have attained.

Surely, the time has now come for a radical re-appraisal of the state of the honeybee in this country and courageous decisions made by competent people in positions of influence to halt and reverse a deteriorating situation. To continue with present policies and attitudes is the road to worse chaos.

Whilst at the time of writing there is no recognition of these problems by our national beekeeping organisations and institutions, or any awareness of the benefits that a National Bee Breeding Programme would bring, there is nevertheless hope for individual beekeepers who are prepared to learn from BIBBA the simple techniques and methods acquired from the German Bee Research Institutes that enable improvements to be made in the bees we already have in this country. The remaining pages of this booklet examine the matter in greater detail and offer guidance to those who are concerned to have better bees.

Chapter 2
The causes and effects of mongrelisation

There is a widespread belief that the honeybees in Britain are a mish-mash of mongrels, many of them being of little merit and incapable of improvement. Recent studies confirm this to be partially but not entirely true.

Before the first known successful importation of foreign bees into Britain in 1859 the British Isles were populated by the Dark European honeybee, which was variously known as the Irish, Welsh, Scottish, English or British Black or Brown bee. Even the scientists caused confusion by their nomenclature. Buttel Reepen gave the name 'lehzeni' to the German Heather bee. The name was later used for the bees of Norway, and by Dr Colin Butler for the bees of Britain, although they are all of the same sub-species.

Some British writers named the native bee *Apis mellifera domestica*, although the hive bee is not domesticated and is indistinguishable from the feral bees. During the remainder of the 19th and early years of the 20th centuries importations of a variety of foreign bees were made, but while these importations may have had some localised effect on the native bee, it was not until the massive imports following the Isle of Wight epidemic that any substantial amount of mongrelisation took place and today it is no doubt true that in many parts of the country mongrel bees prevail.

Mongrelisation occurs as a result of the cross-breeding of imported bees with surviving stocks of the native bee and the subsequent inter-breeding of the cross-breds. Mongrelisation is perpetuated by the continued importation of foreign bees. For completeness one must also state the obvious: foreign bees also cross-breed with mongrels already in existence, as do the native bees.

If a complete ban on foreign bees was imposed, in the course of time the climate would eliminate those already here and ultimately we should find only native and 'near native' types prevailing (see Chapters 6 and 7).

Another effect of the ban would be to concentrate the minds of affected beekeepers to the alternatives. Their attention would be forcibly diverted to the selection and breeding of bees, fully suited to the climate, from the best of those already available.

If, however, foreign bees were to be imported and multiplied on a massive scale in an attempt to displace the native Dark bee, as has been done with the Carniolan in Germany, we should continue to be faced with the problem of mongrelisation since the native bee, sustained by its suitability for the climate, would still survive preferentially in the wild and in the apiaries of those 'let alone' beekeepers who refused to accept the foreign bees.

But there are, regrettably, some people in this country who are awaiting the arrival of *Varroa* in the belief that the import ban will be lifted on bees from Europe. These people should be made fully aware that, whilst hybrids between the Carniolan and the Dark bee are very productive, they are also exceedingly bad tempered (Ruttner (1988), pp 100-101).

In fact, Professor Ruttner tells us these hybrids are not fit to be kept even within the confines of a research institute, let alone by other beekeepers.

Chapter 3
Social problems caused by mongrel bees

1. Bad temper

The experience of beekeepers like ourselves who have kept bees since the war years of 1939-45 (or earlier) testifies to the great deterioration in the temper of bees since that time as also does the prevailing fashion amongst many beekeepers of wearing excessively protective clothing consisting of bee suits, wellington boots and gloves (which make careful handling difficult). These bad tempered bees are a source of nuisance and annoyance to neighbours who in today's consumer conscious society are only too ready to complain or invoke the aid of the law.

To replace bad tempered bees with good tempered bees of foreign origin merely gives temporary, localised relief whilst at the same time adding to the mongrelisation of other bees in the area and thereby making the problem more acute (see Chapter 2).

Professor Ruttner (1988) says: "The more populous our country becomes the more we are restricted in the siting of our apiaries and sites for migratory beekeeping and it becomes correspondingly more important that we pay attention to gentle temper in our bees as the deciding factor in breeding.

Unfortunately, most racial crosses incline towards increased stinginess, even when outstandingly gentle races are used for crossing."

Of the temperament of the Dark bee he says (Ruttner et al, 1990): "The bad reputation of the Dark bee as being aggressive stems, to a great extent according to my experience, from the presence of hybrids in most parts of the native area. The defensive behaviour of unhybridised colonies ranges from docile to rather aggressive. The tendency to sting is highly increased in hybrids with other races; this increase can even be taken as a measure of hybridisation - ('incompatibility of temperament' - Cooper)".

The replacement of bad tempered bees with good tempered bees of native origin provides a potentially permanent solution with no harmful side effects for other beekeepers.

2. Swarms

We are aware that some areas of the country suffer from an excessive number of swarms that are a nuisance to the general public and to the police. Since many beekeepers use swarming as the occasion of re-queening or for increasing the number of colonies, swarming is perpetuated. Again the answer is to produce queens from strains of the native bee that are disinclined to swarm. We are selectively breeding from a supersedure strain of the native Yorkshire bee that, up to now, has very rarely swarmed and is also very gentle, hardy and productive.

There are those who will dispute this statement, but the fact of the matter is that swarminess is an inherited character and young queens produced by a swarmy colony will, most probably, also head swarmy colonies. Since most swarms come from swarmy colonies this character is more prevalent than the supersedure tendency. It is said that supersedure colonies would become extinct if they did not at some time swarm, which some do after four, five or even ten or more years. This is not because of a degeneration in their make-up but because they may raise queens at a time when there is no possibility of the queen being fertilised. Young queens from such colonies produced at the time of swarming would be expected to carry the supersedure character.

We are aware, of course, that with the possibility of multiple mating and the presence of swarmy bees in the same area, supersedure stocks may occasionally produce swarmy offspring. For this reason, constant monitoring and selection are obligatory.

Chapter 4
Economic and management problems

Even when protective clothing is worn, bad tempered bees take longer to deal with than the good tempered.

Swarmy bees require a much higher labour input if swarming is to be controlled, otherwise there is a reduced honey crop. Ruttner (1988, p 52, para 3) states that stocks headed by pure mated queens required 35% less attention, since their uniform development allowed uniform treatment.

The productivity of mongrel bees cannot be improved except by selective breeding over a very long time, therefore it is impracticable for most beekeepers, whilst the productivity of bees of pure race can be improved within a reasonable time (Chapter 5).

The pure native Dark bee requires much less feeding than imported bees or their mongrel descendants, thereby giving a saving of labour and sugar.

It is our experience, though not yet proved by properly controlled scientific tests, that whilst in some cases the imported bee may produce large amounts of honey in hot summers such as 1989 and 1990, the native bee will produce a greater net surplus of honey over a prolonged period of good, indifferent and bad British summers (ie, total amount of honey crop, minus the weight of sugar fed).

Losses of the Dark bee in the exceptionally cold and severe winter of 1985-86 were much lower than those suffered by people who kept imported bees and their mongrels (BIBBA survey 1986, personal communication, A Knight). See Chapter 6 for the reasons.

Chapter 5
The German experience

In Germany, unco-ordinated trials and all possible kinds of imports had resulted in a population of bees that were swarmy, aggressive and unproductive, very similar to the situation in this country at the present time. After the end of World War II a national plan was prepared, based on rigorous selection for pure race breeding, which has produced a remarkably gentle and much more productive bee, which can be relied on to reproduce the desired qualities as long as the programme is adhered to. It happens that this bee is Carniolan because, when the scheme was drawn up, the Carniolan was the only pure race that had been selectively bred and improved that was available. A native of the Danube valley, it was also suited to the Continental climate of Germany.

English visitors to Germany are rightly impressed by the remarkable gentleness of the bees, but make the mistake of thinking that this is merely the attribute of the race and ignore the long process of selection. (It is recognised by some authors that all the four principal European races of the honeybee, Carniolan, Caucasian, Italian and the Dark bee, are gentle when in a pure condition and that hybrids between these gentle races become aggressive, even in the first generation.) At a conference on Bee Breeding held at Lunz in 1985, all speakers agreed that since the introduction of the Breeding Rules of the DIB (The German Beekeepers' Association) the Carniolan bee had improved almost beyond recognition.

What we need to copy from the Germans is the system, not the race to which it has been applied. We can supply a copy, in English, of the DIB Breeding Rules.

Productivity of bees in Germany

The general statement that as much honey is produced in 'West' Germany as in the whole of Germany before 1939 is supported by many detailed reports.
1. At Kirchhain in 1967 stocks from a selected strain averaged almost double the product of unselected bees. Further, in the selected stocks, the poorest yield was 70% of the best, while in the unselected it was only 40%. The greater uniformity of the selected stocks also made management more economic.
2. A breeder, Dr Wohlgemuth, asked his customers about the performance of his queens. Their superiority was almost everywhere confirmed, often producing almost double the yield of unselected stocks.

3. J Dorminger, taking the yield in 1941 as a basis, reported a steady annual increase, over a thirty year period, reaching 190% by 1971.
4. A study of records of testing stations in former West Germany and Austria from 1961 to 1983 showed an average increase of 6.7 kg per colony. This continuous annual increase of 0.25 kg per colony over a 23 year period is the more convincing because it is based on a wide survey: 61,870 Annual Returns were studied.

Full details in Ruttner (1988) pp 50-53.

Pure race breeding

Mendel has shown us what the average results of hybridisation in a large population can be and equally has shown that it is impossible to forecast any individual result, so **Pure Race Breeding** has to be practised, if predictable, desirable results are to be obtained. Ruttner (1988) says: "Experience over many years has shown that lasting results can only be obtained from breeding within a **pure race**, certainly not from breeding with the repeatedly crossed (mongrel) local bees." He also says: "We regard the natural strains and races of bees as a harmoniously balanced system with particular characters and qualities. This system consists of an enormous number of independent single elements which are not linked together but only loosely associated. Every crossing with another race disorganises these associations and creates a multitude of new, mainly less productive, types. The change in the body characters is the outward sign of this disorganisation" (Ruttner, 1988, p 10).

The purity of the race is maintained by constant use of morphometric measurements. These tests are carried out on samples, usually 25 bees, from the intended breeder queens, also from colonies headed by her sisters. Colonies for drone production are checked for the purity and performance of the queen's mother stock, and her sisters. The disciplines were gradually developed from 1916 onwards by years of patient observations in Russia, then in Germany.

There is also a continuous never-ending process of monitoring and recording the gentleness of colonies and the swarming rate, which are the two most important qualities for bees that are to be kept in a heavily populated environment. Productivity is recorded and resistance to disease is looked for.

Although these methods are the end product of the work of scientists in many research institutions, the methods have been simplified so that ordinary beekeepers in Germany (and now some BIBBA members in the British Isles) can undertake the work of morphometric examination and selection.

Danger of relapse

Ruttner (1988, p 103) has written of the chaos which resulted from uncontrolled imports and records the improvements (noted above) in behaviour and productivity which have resulted from Pure Race Breeding, and most emphatically asserts that if once the different disciplines required were relaxed, beekeeping in Germany would relapse into the chaos from which it has been painfully rescued.

The Germans had to replace a native bee by an imported race. Our task should be simpler. Nature would be working with the breeder by favouring the indigenous race which he is trying to improve. In one respect we should be imitating the Austrians who did not have to contend with imports but, for over a century by continuously breeding from selected bees for improvement, at first carried out by practical beekeepers and since about 1946 with the aid of the scientific methods practised in Germany, made possible the improvements noted above.

Racial hybrids

"The innumerable cross-breeding investigations, which have been carried out over the years, have not yet succeeded in discovering racial hybrids which under the conditions of Central Europe were the equals of good pure Carniolan strains. The good performance of racial hybrids is a temporary manifestation which cannot be relied on to reappear in their progeny. Meanwhile, when it is practised, there is uncontrolled hybridisation in the neighbourhood with all its harmful consequences. All the drones that fly out of an apiary influence not only their own, but all the other apiaries within a radius of 10 km." (Ruttner, 1988).

The Buckfast bee is the best known multi-racial hybrid. For an independent appraisal of this bee and the consequences of its use we would refer you to Ruttner (1988), pp 101 - 102.

Chapter 6
The best bee for the British climate

In our earlier years we believed that the grass was greener on the other side of the fence and all the available kinds of imported bees were tried - Italians, Caucasians, Starline, Midnite and Buckfast hybrids and even Anatolians. In each case they were found to be inferior to our local bees. Whilst some of the imports survived and produced honey in the few and infrequent hot summers we sometimes have, they were unsuited to, and many of them were unable to survive, the long spells of cold, wet weather we often have to suffer in many of our summers and the long spells of severe cold that occur in our equally unpredictable winters and springs. The winters of 1946-47 and 1962-63 were, in our experience, equally as bad for bees as was 1985-86. We concluded that we need bees that are able to survive and produce dense, good quality honey regularly year by year in our harsh and variable climate.

We cannot control the climate, but we can breed from bees that are already adapted to it. Such bees are the descendants, not only of those which survived the 'Isle of Wight' disease in the early years of this century, but also of those which have survived in these islands or the nearby continent for the last 10,000 years or so when the re-colonisation by plant and animal life took place after the last Ice Age.

For a fuller understanding of the nature of this subject we need to look back in time to the origins of the geographic races of honeybee that require consideration for our purpose. These are the Dark European, the Italian, the Carniolan and the Caucasian.

Current scientific opinion (see Ruttner, 1988a) is that the Western honeybee probably originated and developed as a successful species (*Apis mellifera*) in the central part of North Africa. It is thought to have spread from this area in three directions: southward, to colonise tropical Africa as far as the Cape, with many sub-species evolving in response to environmental needs; to the east, to colonise the Middle East and south east Europe, evolving into several sub-species of which the best known are the Italian, the Carniolan and the Caucasian; the third migratory route was to the west, across the Sahara which was a savannah before it became a desert, from which evolved the races of North Africa, the Iberian peninsula and, to the north of the Pyrennes and the Alps, the Dark bee.

Repeatedly during the last two million years, northern Europe has been covered with ice or under the immediate influence of the ice cap. During this period the Dark bee was confined to the Mediterranean coastal area of France.

About 10,000 years ago the ice cap began to melt, the climate grew milder and the vegetation of the tundra was replaced by flora which could support bee life and provide suitable nesting places. The Dark bee migrated northwards and colonised the 'British Isles' (at the time not islands but a peninsula from the mainland of Europe) and Europe north of the Alps as far as southern Sweden and the Baltic lands, and eastwards as far as the Urals. New information (Ruttner, Milner and Dews, 1990) reveals that Norway was also colonised at least as far as Oslo.

Recent studies (Ruttner et al, *op cit*) show that, despite extensive importations of bees of other races during the past 150 years into the territory of the Dark bee, colonies of unhybridised Dark bees can still be found in many places. Morphometric studies made at, or in conjunction with the Institute of Bee Research at Oberursel, Germany, confirm the uniformity of samples from the whole of this vast area. As recorded in the next chapter, examination of samples of bees preserved in several museums in Britain, collected before 1859 (the date of the first recorded importation of foreign bees into Britain) and also samples from the Viking excavation in York, *circa* 1000 AD, and from that at Oslo, *circa* 1200 AD, have confirmed the morphometric standards of the Dark bee. Samples from New Zealand and Tasmania, descended from bees taken from this country more than 150 years ago, also conform exactly.

Whilst there is morphological uniformity among the indigenous bees of this vast region, there are physiological and behavioural differences that have developed during the last 10,000 years as the bees became adapted to different environmental demands. The bees that are thus adapted are known as ecotypes. In France at least five of these have been identified, each with its own seasonal brood rhythm. The Russian ecotype of the Dark bee can survive in areas where "the rivers are not frozen over for more than six months of the year." (Alpatov, quoted by Ruttner, 1988a). In Britain, as in other countries bordering the Atlantic or the North Sea, a distinct ecotype has evolved in heather areas that is more inclined to swarm than are bees out of flying distance of the heather. This is a response to the very meagre forage available for most of the year and a sudden abundance for a short period only late in the season.

The foregoing information provides evidence that the ecotype of the Dark bee which, over the past 10,000 years has evolved in any particular area, should be best able to withstand the extremes of climate in that area and, with proper management, be more economically viable than other bees adapted to places with a very different climate.

It is the experience of people who keep the Dark bee in this country that the bee will produce surplus honey every year, even when the summer is so cold and wet that bees of foreign origin have to be fed sugar to keep them alive. This is a consequence of their character of moderate brood production throughout the active season with their compact pattern of brood and always a reserve of stores. A quick reaction of breeding activity to adverse weather conditions results in a low consumption of food. These characters, together with a population of long living worker bees, provide an optimum number of foragers ready to take full advantage of any short nectar flows during periods of unsettled weather. There is at such times a high ratio of potential foragers to brood, in contrast to the more prolific and thriftless Italian bee which continues to maintain a large brood nest in those conditions and also has shorter-lived workers.

The Secretary of the Wakefield and Pontefract BKA has recently examined the reports for the last 20 years. Including the good years of 1989 and 1990, there have been 5 good summers, 3 indifferent, 10 poor and 2 very poor, yet our local native bees have produced surplus honey in every one of these years.

During a considerable part of the last millenium the climate in this country was much more unfavourable to bees than anything we have experienced this century. The Little Ice Age lasted from about 1200 AD to 1850 AD and it is therefore no surprise that the pure form of the native bee came safely through the most recent severe winters of 1946-47, 1962-63 and 1985-86, when losses among foreign bees and their closely related hybrids were very heavy.

The physiological reasons for the survival of the Dark bee in severe winters are given by Ruttner (1988a) :-

1. Efficient thermoregulation of the brood nest.
  a) The Dark bee has the largest body of the whole species with greater metabolic heat production by individual bees when required.
  b) The Dark bee has the longest abdominal overhairs of the European races. The colony forms a 'winter cluster' when the air temperature falls to 2 °C. The bees which form the outer layer tuck their heads inwards and the abdominal overhairs interlock from bee to bee, insulating the cluster like the fur of a mammal.
2. In late summer, perhaps because of the diminution of brood rearing, the amount of biopterin in the larval food is greatly increased and 'winter bees' are formed, in which protein and fat accumulate in the 'fat bodies' in the sub-dermal layers of the abdomen. These bees are still physiologically 'young' in spring and so can act efficiently as nurse bees. It is therefore not necessary to produce brood in the depth of winter in order to have nurse bees in spring, as is the case with Italian bees.
3. There is an increase in the amount of another enzyme, catalase, which enables the rectum to retain greater quantities of faeces during winter. Such bees, confined for long periods in winter without the possibility of a cleansing flight

are less liable to develop dysentery. It has been shown that southern bees taken to a cold climate do not increase their production of catalase.

4. The Dark bee has a longer period without brood in winter and consequently consumes less food, with a reduction in the accumulation of waste products. The more efficient thermoregulation also reduces the intake of food which is needed to maintain temperature within the cluster.

5. The Dark bee has greater resistance to nosema.

Despite all the above facts about the suitability of this bee for our climate and the testimony of those who have had experience of both native and foreign bees, there are still the cynics who pour scorn on the efforts to conserve and improve the native bee. They claim that a 'modern' bee is now needed that can cope with the environmental changes taking place as a result of agricultural practice, eg, the oil seed rape crop, cereal 'deserts' and other major changes that may be imminent. These people should take note that the Dark bee is the most adaptable of all the honeybee races, its territory of natural distribution ranging from the Mediterranean coast of France to Southern Scandinavia (and in the care of man as far north as the Arctic Circle) and from the humid and temperate climate of the Atlantic seaboard of Western Europe to the extremes of severe cold and dry heat of central Russia as far east as the Urals. There are also regional ecotypes within the British Isles with differing patterns of development and behaviour that enable us to choose a bee 'tailor made' for any changes in bee forage either in time or scale that may be imposed on our countryside by economical or political pressures.

It is also claimed that modern beekeeping needs a more prolific bee with a large brood nest. This is based in part on the apparent ability of large colonies to gather a larger crop of honey in a good and sustained nectar flow than smaller colonies. Whilst it is true that a large colony of any one strain of bee will usually produce more honey than a smaller colony of the same strain, it does not necessarily follow that a large colony of a prolific strain with a large brood nest will do better than a strong colony of a less prolific strain with a smaller brood nest. Indeed, as has been mentioned earlier, the opposite is frequently true in the average conditions in this country. Furthermore, it should not be assumed that all the ecotypes of the native bee have small brood nests. Experience indicates some variation, and reference to the writings of Pettigrew and his use of large skeps confirms this.

Chapter 7
The survival of the Dark bee in the British Isles

For many years there has been a widely held belief that the native British bee was virtually exterminated by the Isle of Wight disease, and it was also assumed that any surviving bees had all been mongrelised by the massive imports of foreign bees which were encouraged in the aftermath of the epidemic.

No properly conducted surveys of the whole country were made at the time to substantiate this belief and morphometric techniques were not available to distinguish between the pure Dark bee and the dark coloured mongrel. However, there were many witnesses who testified that their bees survived, the most notable known to us being Robert Couston with his long and wide experience in Scotland.

Morphometric techniques were initiated in Russia from 1916 onwards and were sufficiently developed to be published by Goetze and given practical application from the end of World War II. The practice of morphometry is obligatory for breeders in Germany. The techniques are taught in winter classes to ordinary beekeepers. Till fairly recently the morphometry of the honeybee was unknown in this country and the number of beekeepers who practice it is lamentably few.

Recent morphometric studies have provided convincing evidence that the native Dark bee still exists in a pure state in this country.

In 1988 Professor Ruttner, who knew of the morphometric work we were doing as part of our breeding programme, wrote: "The morphometric standard set down for this bee in Germany is scientifically not sound because of the massive importation of foreign races prior to the start of exact morphometric studies by G Goetze around 1925." To test the force of this objection he asked us if it would be possible to examine museum specimens collected before 1859, the year of the first successful importation of foreign bees into these islands.

Our British museums are much more richly endowed with specimens than was expected. From the oldest specimens in the Natural History Museum (1696) and other collections, we obtained adequate evidence and Dr Ruttner urged us to proceed to publication. Then came the measurement of the York 'Viking' specimens (AD 1000) and the Oslo specimens (AD 1200) and the Linnaean collection in London, which has delayed the completion of the work.

These results, which have created great interest in Germany and Norway, have now been published in The Dark European Honey Bee, *Apis mellifera mellifera Linnaeus* (Ruttner, Milner and Dews, 1990). The results have been summarised by Dr Ruttner in five short statements:

1. **The measurable external characters (the phenotype) of the Dark European honeybee,** *Apis mellifera mellifera L.* have been definitely established by comparing recent samples from NW Europe with British museum specimens collected early in or before the 19th century, and archaeological finds from the excavation of a Viking settlement in York (10th century) and others from Oslo (end of the 12th century).
2. *A m mellifera* **can be significantly differentiated** from its geographic neighbours (the Carnica-Ligustica group) by 12 morphological characters.
3. **The phenotype of the Dark honeybee has not substantially changed,** neither during the last millennium in Europe, nor by transplantation to the southern hemisphere (Tasmania and New Zealand) during the last 150 years.
4. **THE REPEATED CONTENTION THAT THE 'OLD BRITISH BEE' IS EXTINCT CAN BE ASSUMED TO BE DISPROVED.**
5. It is documented that honeybees existed in the Oslo region at the end of the 12th century AD during a warm climatic period, but not later till they were reintroduced in the middle of the 18th century. It can be assumed therefore that the northern limit of *A m mellifera* was shifted southward before 1850 because of deteriorating climate.

The reasons for the survival of the native bee lie in its inherited capacity to thrive in our cool maritime climate, whereas bees of foreign race are lacking in the essential physiological and behavioural characters that would enable them to cope with the climatic extremes of the British Isles. The effects of long periods of confinement in the hive, experienced during the winters of exceptional severity that occur every 15 or 20 years, are well known (Chapter 6).

Dr Ruttner (1990) states that a more significant factor than the cold winter is the failure of young queens to mate in cool and wet summers. This has a strong selective influence in favour of the Dark bee. Beowulf Cooper observed that native queens and drones fly to mate in cooler weather when foreign queens and drones will not leave the hive. An instance of this was reported by Möbus (Ruttner et al, 1990) but more work is required.

It has been shown on the Continent that Carniolan and Italian queens and drones will fly considerable distances in fine weather to mate at drone congregation areas (F and H Ruttner, Koeniger, many published reports in Switzerland and work by Dr J van Praagh) but we are not aware of any studies of matings close to the apiary.

Cooper, however, observed alternative forms of mating behaviour in our native bee which can in simple terms be summarised:
1. **Distant drone congregations** in which, as on the Continent, mating takes place during long spells of hot, settled weather. (Cooper in a lecture at Celle, and also J van Praagh at a conference in Lunz in 1985, quoted Gilbert White of Selbourne as possibly the earliest recorder of this.)
2. **Apiary vicinity mating**, which takes place in brief fine intervals during long periods of cold, wet weather. During such times Cooper also observed that in these conditions a congregation of drones would form in the proximity of the apiary when the shelter of trees or buildings caused a 'bubble' of calm, warm air to develop during relatively short spells of sunshine. It is also suggested that the wing shape and wing muscles of the native bee enable it to fly in stronger winds than bees of foreign origin.

It is our experience that we can obtain a high percentage of pure matings with our strain of native bee in summers such as 1986-1988 when the beekeepers with mongrel or foreign bees had a high failure rate in this district. It is also our experience that in summers such as 1989 and 1990, when there were very long settled periods of hot weather, our native queens mated with Italian drones.

Arising from this problem we have successfully set up an isolated mating station and obtained pure matings from our own drone colonies.

Recent work by Dr Gudrun Koeniger has found that some degree of preferential mating took place when equal numbers of Carniolan and Italian queens and drones were allowed to mate in an isolated drone congregation area. She found that the two races flew to mate at different altitudes and the queens mated selectively with drones of their own race in the ratio of 3:1. Dr Koeniger told us that she believes other factors besides altitude are involved that have not yet been investigated.

So far as we are aware, no research of this kind has been done involving the Dark bee, with the exception of Cooper's observations. There is clearly a need for this work to be done by some properly equipped scientific institution in this country.

Chapter 8
Principles of selection

The reader of text books on bee breeding or bee genetics may become overawed by the apparent difficulties of the practical applications of the subject. On the one hand we are told that if one character alone is selected for, improvement in that character can be readily made, but at the loss or deterioration of other desirable characters, resulting in bees that become, overall, worse than the original stock. On the other hand, if more than one character is to be selected for, the number of colonies required in the breeding programme increases considerably with each additional character so that one soon reaches numbers more associated with the world of astronomy. Yet in spite of this paradox, ordinary beekeepers, with appropriate advice, are successfully breeding better bees.

In this booklet we are trying to set out clearly and simply the essentials for all who wish to see British beekeeping improve, to become more profitable, more pleasant and more socially acceptable. We have been following these methods since 1981 with the direct encouragement of Dr Ruttner and members of the staffs of several beekeeping establishments in Germany. Enthusiastic beekeepers have learned the morphometric techniques and some progress has been made in selecting and propagating native bees which possess the desired characters.

Pursuit of improvement requires widespread, and ultimately nationwide co-operation. Till this is achieved, groups of beekeepers working together can create a zone dominated by good bees in their area and trust to the character of their bees to attract more supporters. The reader who wishes to have improved bees is strongly recommended to study thoroughly Chapter 2 'Selection for Performance' in Ruttner (1988) where this subject is treated in a manner which, for beekeepers in this country, is most revealing. In the meantime we will try to summarise the essential points.

"Breeding is not merely a question of reproduction. Above all, breeding implies improvement in the bee's performance capability. No colony is exactly like another; brood rearing, inclination to swarm, foraging vigour, stinginess, susceptibility to disease, differ from colony to colony. The starting point for selective breeding resides in these differences. Breeding means the augmentation of the best (the positive variants) and the **elimination of the bad** (the negative variants). The aim is the attainment of an apiary which is uniform and has an above average performance."

Breeding does not so much increase the output of the single colony as the average of the entire apiary. For, in apiaries where no breeding is done, a 'Blender' (a colony displaying hybrid vigour) will appear from time to time and give peak results; but together with this there will always be failures, which on the whole bring in nothing, and so considerably depress the total harvest of the apiary.

As a rule uniform progeny, which will breed true, will only be obtained from uniform pure bred colonies. Starting from hybrids or from the uncontrolled mixture of local bees, many years of increase **and** selection would be needed before a stable breeding line of practical use for beekeeping could be obtained. Permanent results can only be achieved in a reasonable time span **within the framework of one of the natural races of honeybees**. Hybrids can only be used as honey getters but must not be bred from.

The starting point in breeding must therefore be the race, that is to say, a combination of genetic qualities, sieved and tested by nature herself. Assessment of physical characters (morphometry) must be made in order to ascertain that this combination has been preserved in its original form and has not been destroyed by crossing. Uniformity of physical characters together with good performance warrants the expectation of uniform progeny. Pure breeding and breeding for production are not at all opposed to each other, as was often asserted in the past, but only the two together provide the basis for successful breeding.

From his breeding strain a breeder requires above all **heritable performance**.

For the critical reader we would point out that the Germans are doing with their bees that which British farmers do with their livestock, ie, they use pure bred pedigree stock. Hybrids are also produced by, or for, our farmers, but their use is for economic purposes only and not for further breeding, and of course this hybrid farm livestock cannot roam around the countryside and breed indiscriminately with other farmers' livestock as our bees do.

Thus there are three aspects to consider in looking for suitable breeding material:
1. Assessment of behavioural characters, as on the BIBBA record card.
2. The evaluation of performance (productivity).
3. The use of morphometry to ensure purity of race and as a check on purity of mating. We would stress that this is its sole purpose. We do not use it to deliberately select for any one physical character as has been done in the past, eg, bees being selectively bred for longer tongues, for this always results in reduced productivity, quite the opposite of what was intended.

It is obvious that selection cannot be made from a single hive. The rules of the DIB require the keeping of 20 stocks before a beekeeper can be recognised as a breeder, and 50 stocks before a group can be recognised. We know from our own experience that raising of queens is difficult without a large backing of stocks in addition to the queen raising stocks. Drone producing stocks are required in even larger numbers, and others to provide young bees for stocking nuclei. This may seem daunting, but from the first we have stressed that beekeepers if they wish to make progress need to work in groups. There are other reasons, such as sharing the work, but the possession of a large force of bees is fundamental.

It has been calculated that, in order to preserve the total number of sex alleles in a population and so minimise the risk of inbreeding, 80 colonies are required. The larger the number of beekeepers who take part in the breeding project the greater the chance of success. Attention must be paid to the rearing of drones as well as queens.

It may be that the number of available colonies within the breeding group make it impossible for Dr Ruttner's advice to be followed in detail regarding the selection of breeder queens by progeny testing (Ruttner (1988), p 59, 'Performance of the Family').

In that case, bearing in mind the essential principles mentioned above, the next best alternative we can recommend is that all available colonies should be assessed for behaviour, and performance, any high performing 'blenders' being identified by morphometry and discarded and then the worst 60% of queens replaced each year by young queens raised from breeder queens in the top 40%. Provided that some control of matings can be achieved (see elsewhere in this book) then improvement should be made.

Before starting assessment of colonies by a breeding group, a list should be made of the characters that are desirable and attainable. We personally prefer bees that are good tempered, disinclined to swarm, winter hardy and thrifty at all times of the year, build up colony strength for the main nectar flows and produce good quality surplus honey even in cold wet summers. As has been already noted, to begin a selection programme to provide such a bee from foundation stock that only possessed one or two of these characters would be a waste of time, but if one is prepared to spend time and effort looking for a strain in the locality that already has most, or all of these qualities, then a breeding programme can be started with confidence of success. We were perhaps fortunate to find a strain of bee with all these qualities, but we have no doubt many such strains can be found in other parts of the country. Our strain was discovered some years ago living in the wild with evidence that many colonies in the same locality had been undisturbed over a long period. Thus natural selection had produced bees that were winter hardy, thrifty, had an annual brood pattern matched to the main nectar flows, disinclined to swarm since they were well out of flying distance of the heather on the moors and very gentle to handle as morphometric examinations reveal no trace of Italian or other foreign influence. So here we have a genetically stable strain of the native Dark bee that merely needs regular monitoring to maintain it in a pure state to retain its inherent good qualities, with the culling from the breeding programme of any mis-mated or otherwise faulty queens.

Queens are never killed because of their age. Since these bees are of a supersedure type and very rarely swarm, the only way of making sure that this character is present is to wait till supersedure preparations begin, usually in the third or fourth year. In 1989 one colony with a 3-year old queen produced as much honey as the other colonies with younger queens and in 1990, at the beginning of her 4th year, the spring honey crop was as good as other colonies. This queen was superseded in July 1990.

Nature, however, does not select bees that produce very high yields of honey, so this is where we, the bee breeders have to take a hand by following the German method of selection for performance (Ruttner (1988), p 59).

Chapter 9
Morphometric identification of the European races of honeybee

The morphometry (measurement of shape) of honeybees is not only neglected in the British Isles, but often derided, and sometimes compared with palmistry, so some general introduction is needed.

By the early 19th century, careful study of animals, with accurate measurements and comparisons, had resulted in greater awareness of the relationship of animals to each other and to their environment. As long ago as 1847 **Bergmann's Rule** was formulated, that *if the distribution of a species of mammal is spread over a large area, it responds with an increase in body size, to conserve heat, in districts where the climate is more severe, whether the increase in severity is caused by movement away from the equator, or by going into higher altitudes.* This rule was supplemented in 1877 by **Allen's Rule** that *the appendages - limbs, tail, ears - grow smaller* for the same reason, heat conservation. It was soon realised that the rules also apply to birds, and later, with modifications, to insects. **Rensch's Rule**, 1929, explains that though the warm parts of a bee's body follow the former rules, the wings, which are not heated, do not grow shorter in more severe climates.

Julian Huxley in 1939 added the concept of a **Cline**: it was observed that certain birds, living in groups on separate islands, differed slightly and had been put in different classes. Huxley suggested re-grouping them in one species, which, spread over a considerable area, showed continuous, regular and measurable changes in response to changes in the environment.

This concept of the Cline has proved very illuminating in studies of the honeybee.

An American, **Ashmead**, in 1904, used the Cubital Index to distinguish the different species of honeybee in India. From 1916 onwards, Russian beekeepers, with the aim of increasing the productivity of their bees, started measuring body size and tongue length. Their methods were improved during the following years till the work was taken up by a German genius, **G Goetze** from about 1925 till his death in 1964. The work was undertaken in other European countries till now it is possible to chart the relationships of all the known honeybees in the world. Morphometry has been developing for more than a century and a half, by work of many scientists of great ability and even genius, and yet it is derided as palmistry by some of our 'experts' in this country.

Should anyone still doubt the value of morphometry, it may be of interest to recall a lecture by Dr Paxton at the 1987 BIBBA Conference at Cardiff, in which he told us of four methods of distinguishing between races of bees. Two were not particularly effective, but the results of DNA analysis and the study of the polymorphism of the enzyme, malate dehydrogenase, were reliable and supported the results of morphometric examination. More recently we have seen extracts from a letter of Professor DR Smith, of the University of Michigan, who has been working on the distribution of the different forms of DNA and the enzyme, in which she states that her work has shown that there are three distinct lines of honeybee:
1. The African (including sub-species like Scutellata, Capensis, Intermissa).
2. The east Mediterranean (including Italian, Carniolan and Caucasian).
3. *A m mellifera*, the West European Dark bee.

"The wide spread West European Dark bee is a line as distinct from other Europeans as either European group is from Africans."

Now whilst these studies of DNA and enzymes are strictly for the scientist, the morphometric techniques we advocate in this book are within the capability of the ordinary beekeeper.

The different races (ie, sub-species) of the Western honeybee (*Apis mellifera*) are identified by a study of some 40 morphological characters.

The Dark European honeybee (*Apis mellifera mellifera*) can be distinguished from the other European honeybees (*A m ligustica, A m carnica, A m caucasica*) by a total of 12 morphological characters, but, for our purposes 3 of these suffice. No other race of *Apis mellifera* has, on the average, a low Cubital Index, negative Discoidal Shift, and long overhair, as the Dark bee has. Where Caucasian influence is suspect it may be necessary to take note of the width of the tomenta and the length of the proboscis.

Methods

The unit of bee life is the colony, not the individual insect, so we try to assess the colony by taking a sample of 20 to 30 bees. The German Institutes seem to be satisfied with 20, the German bee breeders have to take 25, or 50, or 100, according to the purpose. These figures are probably chosen because they easily convert into percentages for the comparison of different colonies and apiaries. The French, from whom we first learned, take 30. As with all living creatures, there is a range of differences within each race and the values of individual bees may overlap with another race, so by taking a number of specimens we can get a view of the norm.

Table for the comparison of the main external characters of the four principal races of European honeybees

	A m mellifera 'Dark' bee	A m ligustica Italian	A m carnica Carniolan	A m caucasica Caucasian*
General appearance	Large, broad short limbs	Medium size, slim, long limbs	Medium size, slim, long limbs	
Body colour – workers	Black, may have small spots on 2nd tergite	1, 2 or 3 <u>yellow</u> rings. Scutellum may also be yellow	Black, may have small spots or 1 ring, leather coloured	Black with perhaps spots or ring
– drones	Dark	Yellow rings	Dark, small spots	
Cubital Index mean	1.7	2.3	2.7	2.0
spread	1.3 - 2.1	2.0 - 2.7	2.4 - 3.0	1.7 - 2.3
CI – drones mean	1.3	1.8	2.0	
spread	1.0 - 1.5	1.6 - 2.0	1.8 - 2.3	
Discoidal Shift	Negative	Positive	Positive	Zero
DS – drones	Negative	Positive	Positive	
Overhairs on 5th tergite	Long 0.4 - 0.6 mm	Short 0.2 - 0.3 mm	Short - medium 0.25 - 0.35 mm	Short 0.25 mm
Hair colour – drones	Brown-black	Yellowish	Grey - browny grey	
Tomentum width on 4th tergite at widest part	Narrow, less than 1/2 of width of tergite, hairs sparse	Broad, more than 1/2 of width of tergite, hairs may be yellowish	Broad, abundant hair, bee looks grey	Broad, plenty of hair, bee looks grey
Proboscis (tongue)	Short 6.0 mm 5.8 - 6.2	Long 6.5 mm 6.3 - 6.6	Long 6.6 mm 6.4 - 6.8	Longest 7.0 mm 6.7 - 7.2

* Information on Caucasians is not readily available.

Taking of samples

For a complete examination, young bees are required. These can be taken by placing a jar over the feed hole, or with a small box from the brood combs. Not more than the necessary number should be taken. If only wings are to be measured, dead bees can be collected from the floorboard in winter.

As soon as possible after collection, the bees should be killed or they will crawl over each other, regurgitating food till it is impossible to measure them. Putting them in the freezer is not satisfactory as it makes the proboscis difficult to measure. Ether or chloroform, if available, are satisfactory, or the sample can be dropped into almost boiling water. This is almost instantaneous and leaves the proboscis relaxed and measurable. These bees will need to be dried and kept in a container that can 'breathe' (as when sending a sample for diagnosis) but not in metal or plastic. The wing measurements are made on the whole sample. The other characters are usually measured on only ten bees. To facilitate the measurement of the tomenta, the abdomen should be stretched by inserting an insect pin under the head, through the thorax till it emerges at the tip of the abdomen.

Measurement of body characters

When available, a microscope with a magnification of x20 or x30, fitted with a measuring graticule, is used for the measurement of the tomenta, overhairs, wing veins and proboscis. Since few beekeepers possess a microscope or the necessary graticules we shall describe methods which are quite adequate with more readily available means. A pair of watchmakers' forceps is very useful. We followed the advice given in Dade and found that forceps from HS Walsh and Sons Ltd, 243, Beckenham Road, Beckenham, Kent, BR3 4TS, were better quality and lower priced than instruments obtained locally. The classifications quoted are those used on the BIBBA Morphometric Record Card. Measurement is of little use unless it is recorded and the records preserved for study and comparison, so writing and recording materials are needed. The back of the BIBBA record card shows how the measurements can be recorded.

A computer programme is now available which, after entering the data of Cubital Index and Discoidal Shift, will present a table with the averages, together with minimum and maximum values, and also produce a histogram of Cubital Index and a scattergram of Cubital Index and Discoidal Shift. Anyone wishing to receive information on this should contact the BIBBA Secretary, Albert Knight.

We have found it desirable to be able to identify each bee in a sample. A simple way is to have a small box divided into compartments which are numbered according to the size of the sample. A bee is placed in each and, after being examined for one character, is returned to the same numbered division. If deviant measurements are found on a number of bees we may assume some cross-breeding. If all the deviant measurements are on one bee it is probably a stray from another hive.

The wing measurements

All the forewings in any sample should be from the same side, either right or left, but not mixed. The reason is that the two sides differ slightly, but in a sample of 20 or more the differences cancel out. The wings are carefully detached from the bodies and arranged in three neat columns of five on a glass slide binder for 35 mm photographic transparencies. Neatness in arrangement will make measurement easier. With care they can be mounted dry, joggling them into place with darning needles held between the thumb and forefinger of each hand. An easier way is to place in a small dish two teaspoonsful of methanol (or methylated spirit, but this spoils the colour of the wings) and into this put two drops of thin sugar syrup.

Each wing is taken in turn in the forceps and the attachment end dipped into the alcohol, the surplus shaken off, and the the wing laid neatly on the slide cover. Take care not to get any fluid on the part to be measured or when dry it will obscure or cause distortion of the part of the wing it covers. When the slide cover is full put on the other half of the slide and mark it at once with the date and number of the stock.

Whether left or right wings are chosen they should be projected for measurement as shown in the diagrams.

The projector should have a good lens, free from distortion. The image is projected on a wall, not a screen. If the wall has a patterned wallpaper it does not matter as a piece of white paper is held on the wall to make the measurements.

The degree of enlargement is not critical, as we are measuring a ratio, not a precise length, but a magnification of about 40 is convenient. Great care should be taken to ensure that the beam from the projector is quite horizontal and at right angles to the wall or the image of the wing will be distorted. Wing measurements can also be made using a microscope with a special measuring graticule, but these are difficult to obtain.

The Cubital Index (CI)

Although wings differ in detail, all honeybee wings show the same pattern (Fig 1).

Along the leading edge of the wing (the top as seen on the screen) there is a long segment which is the **Radial cell** (R). Below this are the three **Cubital cells**, numbered from the wing root, I, II, III. There are two short veins in the third cubital cell marked AB and BC. The length of BC is divided by the length of AB to give the

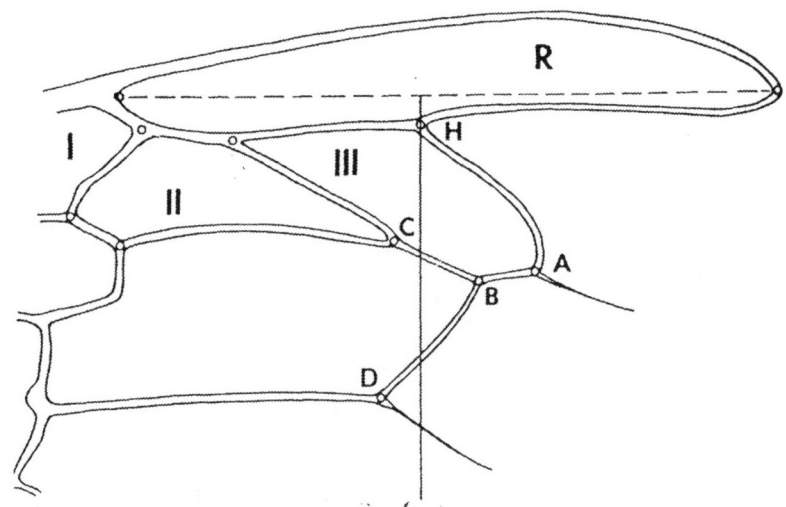

Fig 1 The wing veins used for measuring the Cubital Index and Discoidal Shift
R = Radial cell; I, II, III = Cubital cells; D = Discoidal point

Cubital Index. The measurements must be taken carefully, from the centre of the joints of the veins, and can be done to the nearest half millimetre. When it is known that in every honeybee in the world the angle at this point is 152 degrees, with not more than one degree deviation either way, it will be appreciated that this is a very important and significant part of the wing.

A quicker and easier way is to use the **Index Fan**, devised by **Pfarrer Herold** (Fig 2). (A copy is inserted loose in this booklet.)

It should be held on the wall against the image so that the right vertical passes through the centre of the joint 'A' and the middle vertical through the middle joint 'B' (Fig 3). Great care should be taken in the adjustment so that the vein AB is parallel to the rungs of the ladder. The vein may coincide with a rung, or not, but it must be parallel to the rungs. Then the position of the vein joint 'C' is noted. This is a wide joint and difficult to assess, but efforts should be made to achieve consistency from wing to wing and from sample to sample.

When the centre of the joint has been identified, the eye travels down the fan. The lines are marked at the foot in 0.2 intervals and the Cubital Index should be estimated to the nearest one tenth.

If actual measurements of wing vein length are taken they will have to be recorded

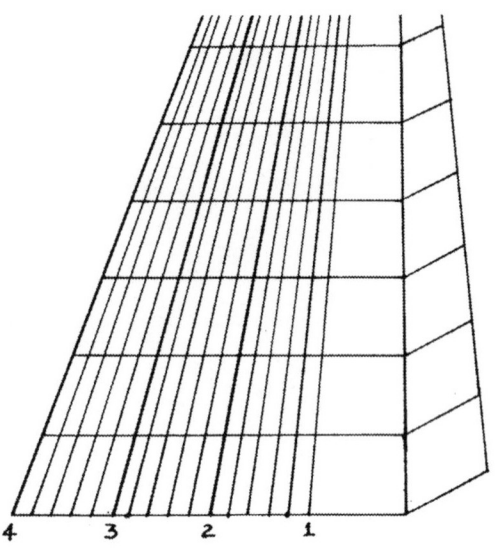

Fig 2   Herold fan for measuring the Cubital Index

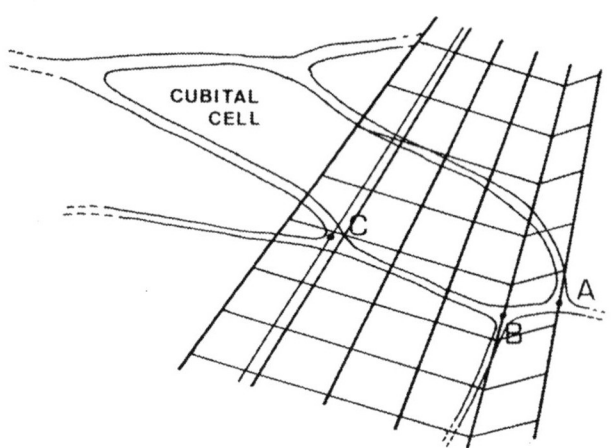

Fig 3   Position of the Herold fan for measuring the Cubital Index

and the Cubital Index calculated. The fan does this part of the work and gives the Cubital Index ready to be recorded. When all the measurements have been made, the mean is calculated by adding all the results and dividing by the number in the sample. The spread (the highest and lowest figures) is also recorded. A pocket calculator is helpful. If the sample is from a Pure Race stock, the figures should be close to those given in the table of characters (p 24).

It should be noted that within any colony the Cubital Index of drones is lower and the Cubital Index of the queen is higher than the Cubital Index of workers.

The Discoidal Shift (DS)

To measure the Discoidal Shift, a piece of white card is marked with a T shape, with four gradations marked on each side of the perpendicular line at 2° intervals (Fig 4). (A copy is to be found, inserted loose in this booklet.)

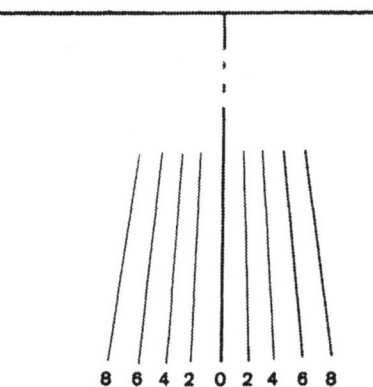

Fig 4   Scale for measuring the Discoidal Shift

This card is held against the wall so that the horizontal line passes through the extremities of the Radial cell and the perpendicular passes through the centre of joint 'H' (Fig 5), which is the junction of the boundary vein of Cubital III with the Radial cell.

The position of the Discoidal joint (D), in respect of the perpendicular is noted. There are three possibilities:

1. Point D may be on the side of the perpendicular nearer the attachment of the wing. It is classed as **Negative**.
2. Point D may lie on the perpendicular. It is classed as **Zero**.
3. Point D may lie on the side towards the tip of the wing. It is classed as **Positive**.

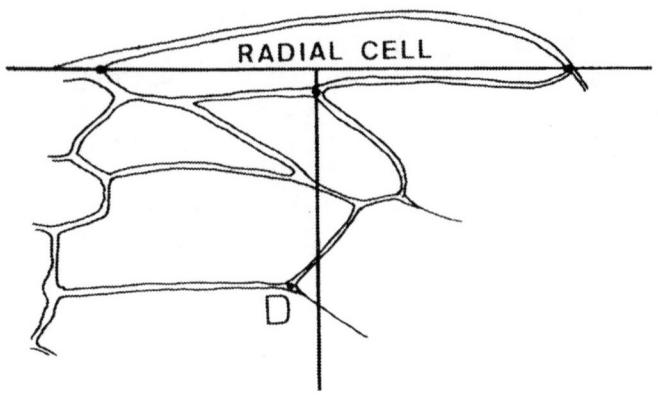

Fig 5   Position of scale for measuring the Discoidal Shift

The measurement is taken at the centre point of the Discoidal joint and each sample is measured to the nearest degree. When the vertical line passes through the middle of the Discoidal joint it is classed as Zero. This joint is about 2° thick, so that when the line passes through the edges of the joint, it is either -1° or +1°.

Within any colony sample, the Discoidal Shift of the queen, workers and drones lies within the same range.

Overhairs

This measurement is second in importance only to the wing measurement. Deviation from the typical length for the race leads to detection of mongrelisation, perhaps from a slight influence several generations back. The hair length is assessed on the 5th tergite, the last which carries a tomentum (Fig 6).

At first it may be difficult to believe that 0.40 of a millimetre is long, but once the hairs of Dark bees and Italian bees have been compared, the difference is striking. The standard of measurement is the first foot-limb width of the back leg of the bee (Fig 7). If the hair is longer than one foot-limb width it is classed as Long, ie, more than

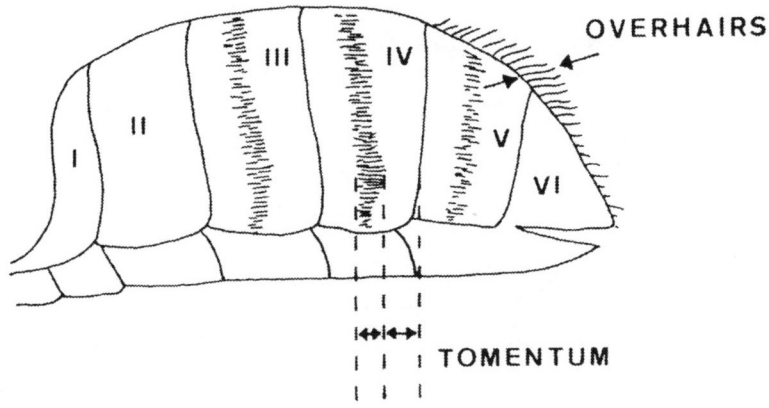

Fig 6   Abdomen of worker bee with numbered tergites, tomenta and overhairs

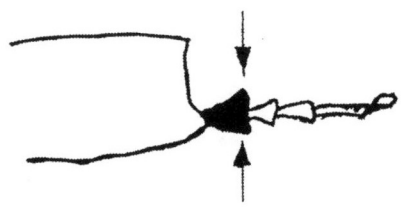

Fig 7   First foot-limb width of hind leg of the bee

0.40 mm, if it is equal it is classed as Medium, 0.35-0.40 mm, and if narrower it is classed as Short, less than 0.35 mm.

Measurements can be made with a microscope, fitted with a suitable graticule. Simpler methods have been devised by German teachers. The main and essential requisite is good light.

Pfarrer Herold's method

A hand lens of x6 or, better, x10, but not more powerful, is needed. To this is taped a piece of wire 0.40 mm calibre, with the end projecting in front and bent at the point of focus of the lens so that it can be seen in sharp outline. The bee is held by the thorax, or impaled on a pin, and approached to the end of the wire till sharp definition is obtained of both the hairs and the wire (Fig 8). With British bees the length of the hair is often very clearly longer than 0.40 mm. With good light and a white background this is quite an easy method.

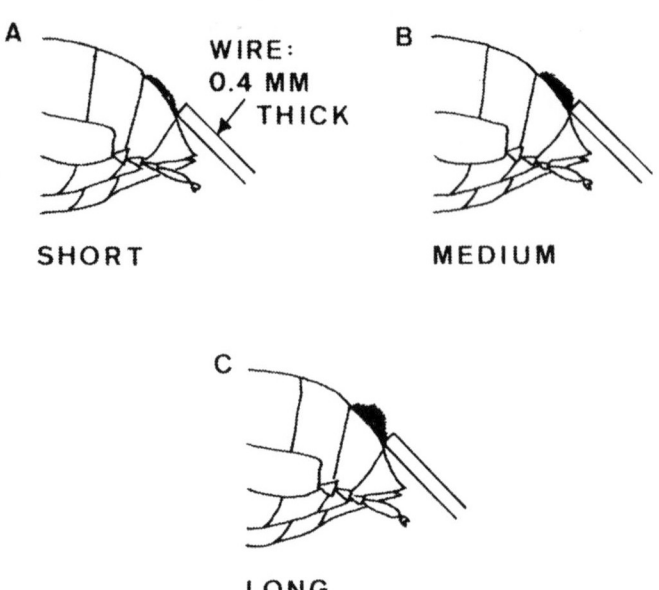

Fig 8   Measurement of the overhairs on the 5th tergite
A piece of wire, 0.4 mm thick, is used to compare with the hair length.
View with a hand lens or watchmaker's eye glass.
(a) Short – overhairs up to 0.35 mm
(b) Medium – overhairs about 0.4 mm
(c) Long – overhairs more than 0.4 mm

Method devised by E Braun, Weingarten

Required: a slide projector from which the slide carrier can be removed (ie, an Aldis type), insect pins and a mounting strip, eg, a strip of cork.

The bees to be examined are impaled through the thorax (upside down, as the image will be inverted) and mounted on the strip of cork which is then inserted in place of the slide carrier. The image can then be focused on the wall, when the hairs should show quite clearly.

The magnification must be determined. Measure, with a micrometer, the thickness of the pin. Focus its image sharply and measure its width to the nearest half millimetre. Divide this by the actual width of the pin to get the magnification. Measure the length of the hairs in the image and divide by the magnification to determine the actual length.

Another way is to make a measuring gauge. The image of a pin or wire of known thickness can be projected on the wall and lines drawn carefully round it on a card. This will represent the thickness of the wire, say 0.40 mm. If this is divided by a parallel line drawn at one eighth of the width of the gauge, a measure of 0.35 mm is given, and at one quarter of the width, 0.30 mm. Another useful method is given in the section on measuring the proboscis.

Body colour

This is easy to determine (Fig 9).

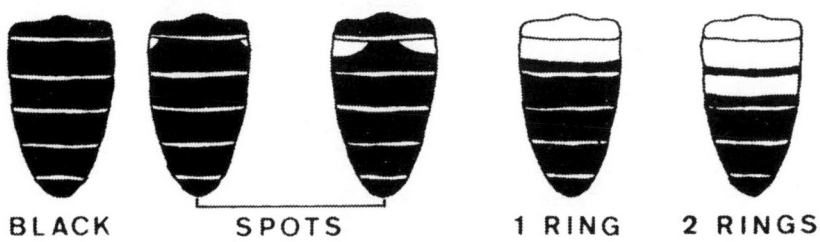

Fig 9   Classification of body colour

Classification is:

Class B   uniformly **dark** in colour. Small brown, tan or leather coloured spots, on the second tergite (scarcely visible to the naked eye) less than 1 mm^2, are ignored

Class S   small spots, larger than 1 mm^2

Class R   the spots have merged to become a coloured ring or band
  R or 1R = one band
  RR or 2R = two bands
  RRR or 3R = three bands

RR and RRR are common in Italian bees or hybrids, and the colour is **yellow**.

The spots or (rarely) rings on Dark bees should be brown, tan or leather coloured, and this is not necessarily a sign of hybridisation.

Tomenta

These are the bands of grey hairs (yellowish in Italians) that cross the abdominal segments, especially on the third, fourth and fifth tergites. The measurement is made on the **fourth tergite**, which is the middle of the three, and usually the widest (Fig 10). The edge is irregular and the measurement is taken at the widest part.

NARROW **MEDIUM** **WIDE**

Fig 10   Classification of tomenta
  (a) Narrow – tomentum less than half the width of the tergite
  (b) Medium – tomentum half the width of the tergite
  (c) Broad – tomentum more than half the width of the tergite

The stretching may have exposed part of the tergite which is not usually visible and this is ignored. The measurement is of the width of the band of hairs and the dark portion towards the tip of the abdomen. If the width of the tomentum is 40% or less of the whole width it is classed as **Narrow**; if between 40% and 50% it is **Medium**; if more than 50% it is **Broad**. This can be estimated with the naked eye, or with the help of a low powered hand lens.

Sometimes the 'Tomentum Index' is mentioned. This requires accurate measurement and is expressed as a ratio, for use in certain kinds of analysis. It does not concern the average beekeeper and is mentioned here to avoid confusion.

Drones have no tomenta, nor are overhairs measured. The general colour of the hair is estimated against the colour scale supplied with 'Breeding Techniques and Selection for Breeding of the Honeybee'. This is not important. Sometimes a band of colour on the posterior edge of the tergite is seen, ranging from scarcely visible to quite distinct, called a Saddle Stripe. It has no known significance.

Proboscis

This is the most difficult part to measure. Take heart! The authors of this booklet taught themselves without instruction or demonstration (fortified by the maxim: What one fool can do, another can do!). The technique is set out in Dade, 'Anatomy and Dissection of the Honeybee', p 106, and the diagrams on Plate 3. If a dissecting microscope is not available, a hand lens mounted on a stand will serve or, as we have found many times in our classes, a sharp eyed person can remove a proboscis neatly without any magnification. It is best to attend classes for this purpose, but for anyone working alone, our advice is to study the diagrams carefully and understand that the proboscis which we are studying is attached to two hinges (cardines in the book – Latin *Cardo*, a hinge) and the upper part, with its two appendages lying close beside it, is covered with a shiny, dark brown membrane with a bulge in the middle which indicates the position of the prementum. Gently slide the tips of the forceps along the side of the prementum till they encounter an obstacle, which will be the lorum or one of the cardines. Secure a grip on this with the tips of the forceps and you should find that the whole of the proboscis comes out in one piece, with the labial palps and maxillary palps adhering. These should be cleaned away (take care! It is very easy to damage the proboscis while you are doing this). The proboscis is placed on a small drop of glycerine on a slide binder.

When all have been excised, the covers are put on, the date and source of the samples recorded. Don't leave it till later.

We are measuring length, not angles or a ratio, so we need to know the magnification. The size of the aperture in the slide mount is nominally 24 x 36 mm, but varies. Measure the width using an engineer's rule, aided by a hand lens, to the nearest tenth of a millimetre. The width of the projected image of the slide frame is then carefully measured to the nearest millimetre and the magnification calculated by dividing this measurement by the actual width of the slide frame. The image of the proboscis is then measured to the nearest millimetre (Fig 11), divided by the magnification, and the result is sufficiently accurate. The use of a pocket calculator helps.

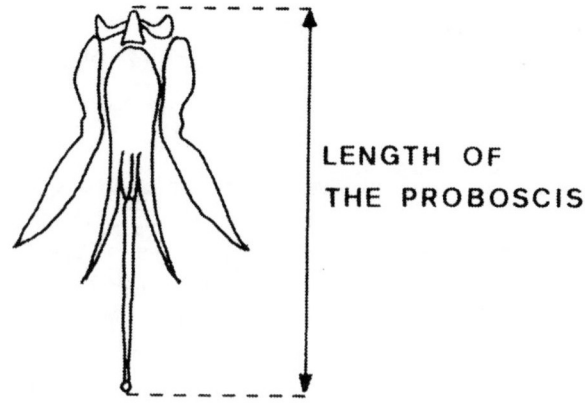

Fig 11   The length of the proboscis

Interpreting the findings

When the measurements of all the characters have been completed and recorded, the results must be assessed. The Cubital Index can be recorded alone on a histogram (a graph in which the values are recorded in columns) (Fig 12). Such a graph can be made as the work proceeds. It is a help if two people work together. Then, while one

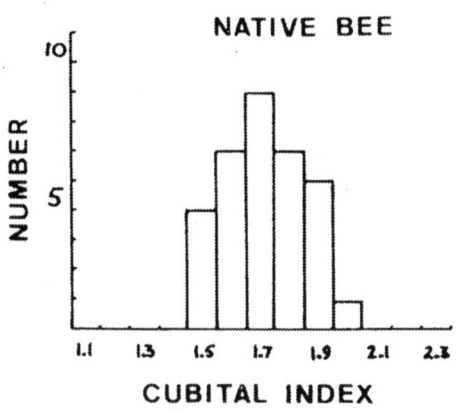

Fig 12   Histogram of the Cubital Index

does the measuring, the other records, and in addition to writing the figures which will be needed to calculate the Mean and the Spread, he can at the same time record, by means of crosses on previously marked squared paper, the numbers in each group of measurements. With a pure bred stock, this should take the form of a single sharp peak, the summit of which will coincide or closely approximate to the calculated Mean, and the extreme points will indicate the spread. Hybridisation will show itself with a flattened peak, or with two or even more minor peaks, so the result is visually apparent.

Scattergram of Cubital Index and Discoidal Shift

This is a method of recording we have devised that shows the relationship of Cubital Index and Discoidal Shift and gives a very clear 'picture' of the degree of racial purity of the sample.

A piece of graph paper is marked out as shown in Fig 13, which is the same as shown on the back of the BIBBA Record Card. Discoidal Shift is recorded in the columns of squares from the base of the scattergram. The three central vertical columns record the bees in the zero category, ie, -1°, 0 or +1°. Negative Discoidal Shift is recorded to the left of these, 1° for each column, and positive Discoidal Shift to the right of the zero columns. Perhaps we should explain at this point why there are three columns allocated to Zero on the scattergram. In our earlier studies, we were comparing our results with the work of M Louis in France. He recorded every wing as zero when the line passed through any part of the Discoidal joint. All other wings were simply classed as negative or positive. He did not measure in degrees. Having previously published scattergrams in this format, it was decided not to change for this booklet. Thus, as we explained on page 30, our measurements of -1° or +1° would all have been recorded as zero by Louis. Cubital Index is recorded in the horizontal lines of squares.

A thick line is drawn across the scattergram between the rows of squares for a Cubital Index of 2.1 and 2.2, indicating the limits of the values for the Dark bee.

For the benefit of those readers who left school before 'modern' maths were taught, we will explain how to use the scattergram. As an example, we will take a bee with a Cubital Index of 1.8 and Discoidal Shift of -5° (Fig 13). Look along the base line and then follow the column of squares marked -5 upwards until you find the square that is opposite 1.8. Mark a small dot in that square. Repeat the process for the remainder of the sample. A fine fibre tip pen, eg, Pilot Hi-Tecpoint V5, is suitable for this purpose. You will now have a pattern of dots. A tightly packed, roughly circular pattern indicates a pure bred or genetically stable bee. A widely spread pattern indicates a cross-bred or hybrid bee.

In Fig 14, sample B8 is a pure bred Dark bee, as all the spots are below the 2.1 line and do not extend into the positive area. If more than 10% of a sample extends above the 2.1 line or into the positive area, then this indicates some degree of hybridisation. Sample B17 is a cross-bred colony with preponderance of Dark bee influence. Sample 22 is a bee of foreign origin. These three samples were all in the same apiary. B8 was the only one to survive the winter of 1985-86.

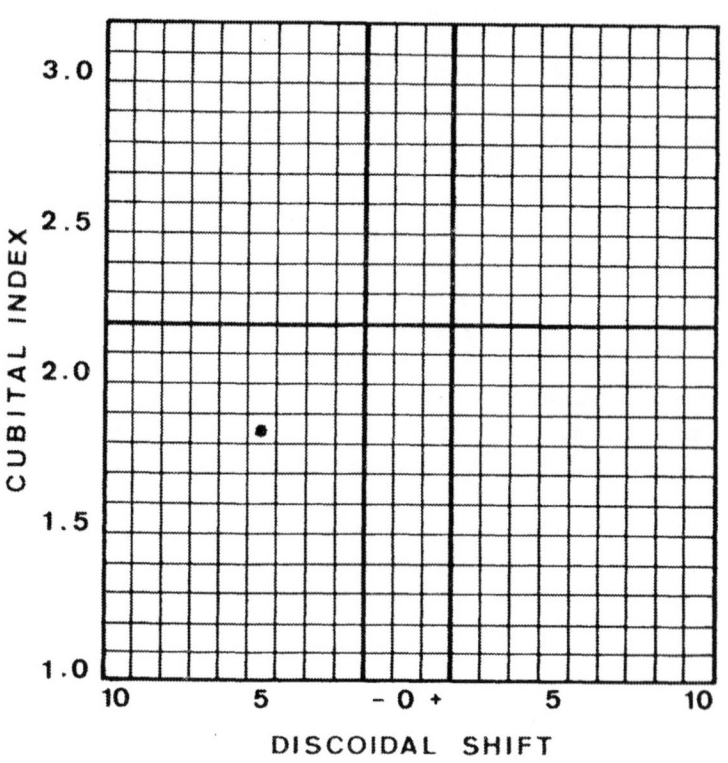

Fig 13 Use of a scattergram showing the point for a single bee with Cubital Index = 1.8 and Discoidal Shift = −5°

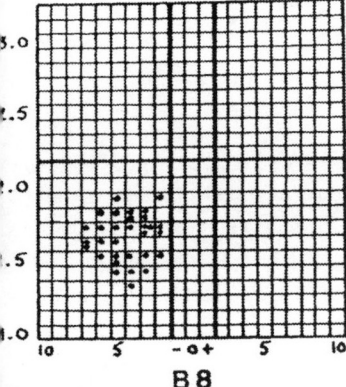
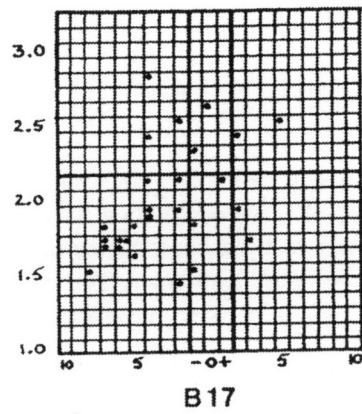
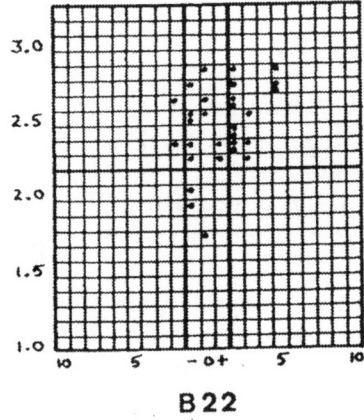

Fig 14 Scattergrams of Discoidal Shift and Cubital Index; each point represents one bee:
Colony B8: low Cubital Index and negative Discoidal Shift — Dark Bee
Colony B17: points over all fields; hybrid with predominance of 'dark'
Colony B22: Cubital Index high, Discoidal Shift 0 and positive — A m carnica or A m ligustica type
All colonies can be easily discriminated morphometrically

Chapter 10
Queen cell production in a single brood chamber colony

The following simple method was observed by BIBBA's Director, Ken Ibbotson, in Holland. It is suitable for a beekeeper with single brood chamber colonies wishing to rear a small number of queens. It should produce 7-10 very good queen cells.

For those with double brood chamber colonies, Method II, detailed in Ruttner (1988), p 26, is recommended. (Note that here the 'honey chamber' refers to the second brood chamber.)

There are three principles involved:
1. A queenless colony will produce queen cells.
2. A queenright colony will complete 'started' queen cells, if these are protected from access by the queen.
3. The colony used to produce the queen cells should not be making preparations to swarm.

Equipment required

1 nucleus box containing two dummy frames.
1 shallow comb with queen cell cups attached along the bottom edge.
1 'Dutch Cage'. This is a cage made of queen excluder which fits over the shallow comb and makes it up to the size of a brood comb.

Step 1

The comb on which the queen is found is transferred to a nucleus box, complete with the queen and the bees. It is placed between the two dummy frames.

A gap is left in the cell raising colony between two combs of unsealed brood.

Step 2

After 1 hour and up to a maximum of 3 hours after the queen has been removed, larvae from the desired breeder queen are grafted into the queen cell cups on the bottom of the shallow comb and this is inserted into the gap in the cell raising colony.

The timing here is fairly critical. If the grafted larvae are inserted too soon after the queen has been removed, the bees will eat them. If they are inserted too late, the bees will have started emergency queen cells on their own brood.

Step 3

Twenty-four hours after the queen was removed, the 'Dutch Cage' is fitted over the shallow frame and it is replaced in the cell raising colony. The queen and her comb of bees are then returned to the colony.

It should be noted that any queen cells that have been started by another queenless colony for 1-2 days can be inserted into a queenright colony if they are protected by a 'Dutch Cage'.

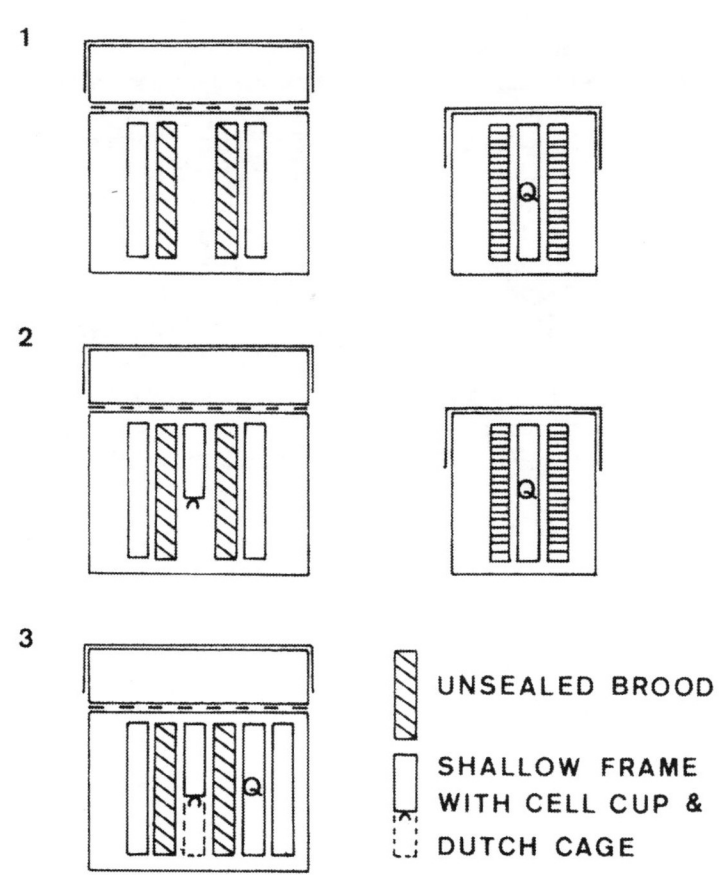

The timetable

The timetable can be summarised as follows:

Day 1   Remove the queen from the cell raising colony, and the frame of bees on which she is found, into a nucleus box.
Insert the shallow comb with larvae grafted into queen cups along the bottom of the frame.

Day 2   Place the shallow comb and accepted larvae into the 'Dutch Cage' and return it to the cell raising colony.
Return the queen and bees to the cell raising colony.

Day 6* The queen cells are sealed.

Day 8   Make up sufficient three-comb mating nuclei to take the sealed queen cells (see Chapter 11).

Day 11 Introduce the sealed queen cells into the mating nuclei in aluminium foil protectors, which have a hole at the tip of the cell to allow the queen to emerge.

Day 13* The queens hatch.

Day 34 Examine the mating nuclei for the presence of a properly mated queen.

* These timings may vary by 1 day, depending on the age of the larvae grafted.

The nuclei should not be disturbed between days 11 and 34 except to feed if required or to deal with robbing. Disturbance, especially around mid-day, could lead to the queen being lost when returning from her mating flight, or being balled by the disturbed bees in the hive.

Chapter 11
Making up and use of three-frame nuclei for queen mating and introduction

The use of a standard frame nucleus is recommended for mating a small number of queens. It can help overcome the problems encountered when introducing the new queen to another colony, particularly if the two are of different races.

For those wishing to raise a large number of queens, the use of mini-nucs is recommended.

A mating nucleus should contain sufficient food and bees to enable the small colony to survive without interference until the queen is mated and laying. It should consist of two combs of honey and pollen stores and 1 comb of bees and sealed and emerging brood.

The nucleus box itself should have a small entrance which can easily be defended by the bees. A half-inch diameter hole is quite large enough. This should be made about 2 inches below the frame tops in the front of the hive, preferably hidden from view by the roof. The bottom entrance, if any, should be closed. Ventilation holes should be cut in the floor and covered with perforated zinc.

If the mating nucleus is to remain at the site where it was made up, the bees from a second comb of unsealed brood should be shaken into it to compensate for the subsequent loss of flying bees returning to their original colony.

Mating nuclei are best placed in the apiary where their entrances are concealed, eg, facing into the bottom of a hedge or in long grass. This helps to avoid robbing by other colonies.

Causes of failure of three-frame mating nuclei

1. The nucleus is made up too late, too close to the time when the queen cell is introduced.
   Make up mating nuclei seven days after grafting the day-old larvae.
2. There are too few nurse bees in the mating nucleus to care for the brood.
   Shake in extra bees from combs of unsealed brood, particularly if the nucleus is to stay at the same apiary in which it was made up.
3. The mating nucleus is robbed out.
   If possible, move the mating nucleus to a different site. Give it a small, concealed entrance and position it facing into a hedge or down in long grass to hide the

entrance from robber bees or wasps (particularly for matings late in the season). Ensure the nucleus has enough bees in it to defend its entrance.
4. The mating nucleus dies of starvation.
Make sure that the mating nucleus contains two combs of food when it is made up. If these are not available, or, if the colony becomes short of food, feed it with candy. To reduce the chances of robbing, begin feeding late in the evening and do not feed the nucleus within 24 hours of making it up.
5. Interference by the beekeeper can cause the loss of the queen. If the mating nucleus is open when the virgin queen returns from her mating flight, she may well be confused and not find her way back to the hive. This is particularly so around mid-day/early afternoon. Disturbances by the beekeeper may well cause the bees to ball the queen and kill her. If the beekeeper keeps the mating nucleus open too long, this could give rise to robbing from other colonies.
Do not look through the mating nucleus for at least three weeks after introducing the queen cell, however curious you may be!

Queen introduction using a three-frame nucleus

"I never feel confident that the queen is accepted until she is surrounded by her own progeny." Brother Adam.

"Introducing a queen of a different race is very difficult." Professor Ruttner, Dr J van Praagh, Dr F Schaper.

When a queen is raised in a three-frame nucleus and remains there to build this into a full sized colony, there is no serious problem. However, if it is intended to remove her and introduce her to a full colony, there is a danger that she will be killed or replaced soon after introduction.

In our early days, we found that even the contents of a mini nuc were not sufficient protection for a queen on introduction to another colony. At Lunz, we were advised to use the following method.

1. Prepare an empty brood box to receive two combs of food and a comb of emerging brood, **without any bees**.
2. Remove such a comb from the stock which is to receive the new queen. Close the gap in this colony either by inserting a fresh comb or moving the remainder.
3. Cover the brood box with a double, bee-proof screen, and place the new brood box above, with the new queen and her attendants, if any. Give this top brood box a small entrance (to guard against robbing) facing towards the back of the hive.

4. The entrance should be kept closed for a few days until there are sufficient bees inside to defend it. The young bees, having never known another queen, will accept and care for the new queen.
   If desired, a succession of combs, always without bees, can be raised from the lower to the upper brood box. As two queens will be laying, the colony will build up more rapidly, but it will take longer to change its character.
5. In due course, when the new queen is well established in the top brood box, the old queen can be removed and the two colonies united by the newspaper method.

This method may seem tedious, but it works, and is better than a quick way which loses the queen.

Chapter 12
The organisation of a local bee breeding programme

Another BIBBA publication 'Guidelines for Bee Breeding' gives advice on this subject and has several appendices which deal with some specific problems based on practical experience in this country. The following information is intended to be complementary to the Guidelines booklet.
1. It is assumed that a number of beekeepers are enthusiastic enough to work together with the aim of producing better bees, with the qualities needed for worthwhile beekeeping in a crowded environment.
2. The next requirement is to find bees that are worth breeding from. Look round among your own bees, or among those belonging to old beekeepers who have not succumbed to advertisements. Good bees have been found in the most unlikely places. Native bees that have not been interfered with can enjoy 'ecological isolation' and sometimes breed true even when surrounded by foreigners. All should be agreed on the required qualities (Chapter 8), especially good temper.
3. If you cannot find suitable breeding material in your neighbourhood, look elsewhere.
a) Obtain suitable breeding material from a reliable source.
Probably the best way is to obtain larvae (the Jenter method has proved reliable) which will enable you to raise a large number of queens at little cost. These queens will then mate locally, probably give good honey yields with perhaps some deterioration of temper, but, most important of all, will produce **pure bred drones**. These drones will have come from one mother, limiting the variety of sex alleles present, and cause inbreeding, which however should be easily remedied later. Different methods of introducing new material are dealt with below.
b) The same autumn, in preparation for next year, try to arrange to have a comb with a small patch of drone cells near the centre of the broodnest, which will encourage early production of drones and, early in the spring, put a drone rearing comb near the brood nest. Drone rearing combs are best produced during a time of colony prosperity from either a frame of drone foundation or an empty wired frame with a starter, which need not be drone foundation. The bees will soon fill

it with drone comb. If you have not raised enough queens to requeen all the hives of the Breeding Group, the remainder can be brought into the scheme as drone producing colonies by the gift of a similar comb.

Before the drones in this frame emerge, remove the comb and replace it with a comb from one of the breeder queens colonies in which the eggs have just hatched. Bees will accept drone larvae but <u>not eggs</u>. As the drones take 24 days to develop, this procedure will need to be repeated about every three weeks. You will thus greatly increase the number of desirable drones and diminish the numbers of the undesirable. The combs with 'bad' drones can be left exposed for the birds to clean and then put once more into a breeding hive.

During the second year of the scheme, obtain more larvae from another good source. These should come from a strain with similar morphometric and behavioural characters. The queens raised from these will have a better chance of pure mating. Multiple matings take place and in your early days it may be necessary to breed from queens that show a proportion of foreign types in their progeny.

A very simple way of improving your strain at this stage is to eliminate any light coloured queens from those you raise and any that produce light coloured drones. You can expect to keep on improving your stock. If, because of fine hot weather during the breeding season, cross matings continue, it may be necessary to continue introducing more breeding material until genetic stability has been achieved.

Displacement breeding

This is a deliberate policy to change the bee population of your district by persuading your neighbours to follow your example on a definite plan, so that your original base is surrounded by an increasing area where only the kind of drones you wish to breed from exist. If at demonstrations and summer meetings your neighbours see the good qualities of your bees, they will be more ready to join. Handling good tempered bees without smoke and with bare hands can make more impression than a course of lectures.

Displacement breeding requires the production of as many queens as possible in an attempt to establish a 'pure race' or 'monostrain' area and this is best done by the method described by Ruttner (1988), pp 74-78. Pure mated queens of the selected strain are produced and tested by the breeding group or chief bee breeder (who needs more skills than a mere queen raiser). Then breeding material is supplied by the breeding group or breeder to as many beekeepers in the area as possible.

The queens thus obtained are apiary mated with local drones and produce hybrid workers. These are known as 'working queens' or 'production colonies'. They function as honey producing colonies and **produce pure bred drones**. They should not be used for further queen rearing.

In the following year, more pure bred queens are raised and allowed to mate locally. Morphometric tests will show if they have been pure mated, that the whole area is saturated with drones of pure race so the production colonies are also of pure race. To raise queens from these colonies before this condition is attained will lead to disappointment and the production of more mongrels.

Some ways of achieving pure matings of native bees

1.　Areas where pure native bees have been identified

We get a good number of pure matings by taking mininucs to the apiary where the strain originated, which is in an area with few beekeepers. If it is known that there are other bees of a different race in the area, then queen rearing should be done at the time of natural queen renewal of the local native bee. This can be ascertained by regular inspections, through the season, of native colonies with marked queens, ie, take note when supersedure takes place that results in a pure mating, or swarming preparations are made by the bees that are not influenced by management of the beekeeper. Successful supersedure can take place at any time during the active season but has been known to happen from very early (late March to early May) to very late (August - September). Matings that take place during a settled spell of hot weather are more likely to be with drones from further away and with drones of a different race.

2.　Areas where no native bees are to be found

Instrumental insemination can be considered. The equipment is expensive, the operation difficult but, with care and a full knowledge of the problems, can be successful, but experience so far in this country indicates that the operator needs regular and extensive practice.

Natural matings will be more likely to succeed if a large number of drones is produced and maintained in good condition and in a place where a 'bubble assembly' is likely to form (Cooper, p 68) and queen mating takes place at a time when there will be few foreign drones flying:-
  a. Very early in the season, but this has the risk of drones from early swarming colonies, especially during the rape flow.
  b. Very late in the season; this has problems which can be overcome, ie, production of viable drones, and queen cell raising is difficult.

c. During cool, wet and/or windy weather.
d. By keeping the queens and selected drones confined in a cool dark place during the day and releasing them each day after normal drone flight has ceased. A report in the British Bee Journal, May 1989, p 99, 'Time isolation for mating', indicates that this is successful in Poland. In 1992 we obtained mated queens by this method. The full details are published in the Spring 1993 issue of 'The Bee Breeder' (BIBBA).

The purity of a mating can be checked from the Cubital Index and the Discoidal Shift of a sample of bees from the early progeny of the newly mated queen.

Distribution of breeding material

1. Mated queens

Pure bred native queens produced by the breeding group are quite likely to be killed or superseded during the first month when introduced to a colony of mongrel or foreign bees. This is a waste of valuable resources. If mated queens are to be distributed, they should be preferably well established in a strong nucleus hive and stocked entirely by the queen's own workers. This limits the number of queens for distribution.

A method for queen introduction can be found on pages 44-45.

2. Ripe queen cells (ie, within 24 hours of emergence)

These are distributed to be put in 3-comb nucleus hives that should have **been made up three days earlier**, containing 1 comb of brood, 2 combs of food and sufficient bees. Sealed queen cells should not be handled until the day before emergence (see Ruttner (1988) for details).

3. Virgin queens

These, preferably from an incubator, are given to **broodless** mininucs (see Möbus: Mating in Miniature).

4. Larvae in Jenter plugs

Such larvae can be supplied to a beekeeper or group that is capable of raising **good quality queens**. This is the most effective way of producing a large number of queens with minimum effort by the breeder or breeding group. Such a person or group can also assist in methods 2, 3 and 5.

5. **Larvae in Jenter plugs that have spent 24 hours in a cell-raising colony**

The 'started cells can be placed in well stocked nucleus hives that have a comb of sealed brood and two combs of food. We have not tried this method, but if only 1 or 2 started cells are placed in such a nucleus hive, good quality queens ought to be produced.

6. **A method described by John Cox (Gloucestershire CBI) at a BIBBA conference**

A small piece of comb, about 1 inch square, with eggs and very young larvae, is cut out of a comb in a breeder colony and inserted into a matching hole cut in an empty brood comb. This is given to a **strong** nucleus, well stocked by nurse bees and with two other combs of food. As there will be only about 30 larvae to feed, a good quality queen should be produced.

Methods 2 and 4 are recommended and should lead to a better acceptance and survival of the queens.

Robbing. These small nuclei are very vulnerable. Watch should be kept and care be taken and robbing screens, described by Steve Taber in 'Breeding Super Bees', used if necessary.

It is most important that all who practise any one of these methods **adhere to the timetable**, which is dictated by the rate of development of the bees. We heard of one person, supposedly working Method 6, who was busy and did not insert the eggs into the hive for several days. Others, with Method 2, did not make up the nuclei till the queen cell was ready to be given and the bees wee not yet in a receptive mood.

A brief note on the management of the native bee

The reader who is sufficiently convinced to change to the native bee should realise that the system of management must be adapted to meet the needs of the bee and not the other way round. As mentioned earlier, there are many different ecotypes of the native bee surviving, each with its own behavioural pattern suited to its own environmental conditions. Note must be taken of this and an appropriate method of management adopted. Supersedure strains of bee will thrive in many places with a system of minimum interference by the beekeeper, whilst the energy generated by the natural swarming tendency of the heather bee should be harnessed to a system of management that results in a loarge force of foraging bees headed by a young queen ready to make the most of the heather harvest. Above all, one should realise that many management methods advocated in our English text books result from the author's experience with imported bees and are unsuited to the needs of the native bee.

For more information on the behaviour and needs of the native bee we suggest you study 'Honeybees of the British Isles', by Cooper, and also 'The Dark European Honey Bee', by Ruttner, Milner and Dews.

Chapter 13
Varroa and the bee breeder

At the time of preparing a revised edition of this booklet (March 1993), the 'establishment' view is that *Varroa* is here, it is going to be with us for evermore and the only option is to control it by chemical or management methods which are expensive in materials or labour, or hazardous in use. It seems to us that an acceptance of this situation is defeatist, unimaginative and scientifically narrow.

As bee breeders, we should be taking a broader scientific view based on a consideration of the relationship of pathogens to their hosts. In a talk on BBC Radio 3 in 1992, an eminent scientist declared that it is not nature's way for a pathogen to exterminate the entire population of its host, since it would thereby ultimately destroy itself. A balance usually occurs when a resistant strain evolves in the host population that ensures the survival of the species.

When *Varroa* was first discovered in England, we considered that, since it was not a natural pathogen of *Apis mellifera*, then it was somewhat unlikely that resistant bees would evolve. However, we recalled that myxomatosis, a virus introduced into this country by man, did not destroy the total rabbit population. Resistant strains soon began to breed and multiply.

We will not go into the controversial question of whether Acarine was the sole or principal cause of the heavy losses of bees in this country in the early years of this century, but it is interesting to note the assertion of Professor Morse (personal communication) that since Acarine mites were first discovered in Texas in 1984, some 90% of colonies in the USA have perished, but colony numbers have been largely maintained by splitting surviving stocks. Whilst Acarine still kills some colonies in this country, most beekeepers are unaware of its presence.

Taking note of MAFF statistics (1986), it is reasonable to assume that there are probably about 250,000 colonies of bees on mainland Britain. If only a few of these colonies are resistant to, or tolerant of *Varroa*, and one of these can be identified, then we have the means of breeding a bee that requires either no treatment, or at any rate, less treatment than is necessary at present. In this we are encouraged by the report from Tunisia, where little or no treatment was given, of the emergence of resistant bees. In evolutionary and morphological terms, these bees, *Apis mellifera intermissa*, are more closely related to our bees, *A m mellifera*, than either *A m carnica* or *A m ligustica*.

We sent the preceding notes on an alternative approach to *Varroa* to some of the leading authorities on bee breeding in Europe and the USA. We are greatly encouraged by their response.

Professor Roger Morse kindly sent an account, published in the *American Bee Journal*, July 1991, 433-434, of the discovery of the first colony resistant to *Varroa* found in Florida in March 1990. This colony, described as 'prosperous', was the only colony alive in an apiary where approximately 60 other colonies were dead from *Varroa* mites. It is estimated that probably some 25,000 colonies in the surrounding area also died from *Varroa*. The resistance mechanism is much the same as that of *Apis cerana*. Dead *Varroa* mites found on the hive floor had indentations on either the dorsal surface of the body or mutilation of the ventral surface. A small number of mites had part of the dorsal surface torn away. It is assumed that a combination of grooming and biting by the bees allows the resistant colony to keep the *Varroa* mite level low.

Professor Friedrich Ruttner informs us that in Austria, a commercial beekeeper, Alois Wallner, used a low efficiency treatment (formic acid) on his 700 colonies, which resulted in a high infestation rate and some losses. However, he found that 12 of his colonies had a lower infestation level and also a higher rate of damaged mites which had legs bitten off by the bees. By requeening his colonies from these resistant stocks, he has not had to give his colonies treatment of any kind for the past three years.

Professor Ruttner says that present experience indicates that colonies survive without any treatment when 60% of dead mites on the floorboard are damaged by the bees, and that when a level of 45% of damaged mites is found, this can be increased to 60-70% by selection, within a few generations.

Other resistance mechanisms have been observed, but the results of grooming by the bees are most easily observed by beekeepers.

The most obvious way of detecting resistant bees requires the monitoring of colonies receiving no treatment of any kind, although losses would be very heavy. At the moment, it seems to us that the only bees in this country that will not receive medication or manipulative treatment will be those living in the wild. If wild colonies can be monitored for continued survival, and not their demise and replacement by a swarm from a treated colony, then we might, somewhere, find resistant colonies. The problem here lies in knowing of the existence of a sufficiently large number of wild colonies that can be monitored. To be effective, this approach would need the creation of many more 'wild' colonies.

But if we are to find resistant bees in this country, we shall have to devise a policy that achieves a balance between that which is biologically desirable, ie, no treatment, and that which is economically and politically acceptable.

We believe that an acceptably effective course of action would be to:
1. Use floorboard inserts to ascertain the level of infestation and also (when no diagnostic chemicals are used) the percentage of damaged mites.
2. Keep susceptible colonies alive by some appropriate treatment until they can be requeened from resistant colonies.
3. Colonies that show the possibility of resistance should be maintained by 'soft' treatments that keep the level of infestation high, but below the level that would kill the colony. These colonies should be 'closed-bred' to intensify the resistance mechanism.

If this policy is adopted by a sufficiently large number of beekeepers, there is good reason to hope that the menace of *Varroa* can be reduced to the same nuisance level as Acarine.

Chapter 14
Measuring the Cubital Index and Discoidal Shift with the aid of a computer program

When using a computer for morphometry the wings are mounted in GEPE double-glass slides as for the manual method, Thin strips of double-sided sellotape 5mm wide are used to hold the wings in position, these are positioned on the slide as follows, one on the extreme left hand side, one 10mm to the right of that and another 10mm again to the right of the second one. Using a pair of tweezers the wings are placed with the wing root on the sellotape strips, 5 in each vertical column, keeping them as neatly horizontal as possible. Normally 30 wings are used for each colony sample, so two slides are required for each sample.

Alternatively the wings can be mounted on a sheet of acetate film as used for OHP's. and scanned using a flat bed scanner with a resolution of 1200 d.p.i.

The slides are scanned in a film scanner and the scanned images saved as files. Both the computer programs described below use this method although the Beewings program needs each wing scanned and saved as a file, whereas the Beemorph program uses a whole batch of wings saved as a file which saves time in scanning.

These computer programs are designed to assist in the measurement of wing indices and also to present the data in the form of graphs. They are valuable aids to recognising if the colony that has been sampled is a hybrid of one or more sub-species, or if it is a pure-bred colony belonging to a particular sub-species.

The Beemorph, designed by Russell Talbot, is available on the Internet free of charge for a 30 day trial, thereafter one can become a registered user for £25.

Russell's web site is: http://www.hockerley.plus.com

A feature of this program is that instead of having a file for each wing, one can have 15 wings or 30 wings in one file, thus cutting down the scanning work considerably.

In operation there are four panels at the same time on the screen, the **Wing Panel** shows the wing that is being worked on, another is a **Spreadsheet Panel** with a tool bar at the top, another is a **Navigation panel** that indicates which of the wings is being worked on, and finally there is a **Guide panel** that indicates the next point on the wing that has to be clicked on.

A sample of wings will earlier have been mounted and scanned in a scanner using a resolution of at least 1200 dpi. The wings can be mounted either in transparency slides - 15 wings per slide - and scanned using a film scanner; or mounted on an acetate film such as used in overhead projectors and scanned using a flat bed scanner.

To start one loads the program, this shows a panel in the top left-hand corner of the screen (Fig 15) The tool bar at the top of the panel has various icons indicating the tools available. From left to right, the **Wing Icon** opens the browser to load the file of wings, the sixth icon across puts the **Guide Icon** on the screen and the seventh icon across puts the **Navigation Panel** on the screen.

Once the file has been loaded the first wing in the file will appear in the lower half of the screen, click on the first joint as indicated by the Guide Panel. This will put two co-ordinate values into the spreadsheet. Once all 7 points have been clicked on, a prompt will appear in the **Guide Panel** for a new wing to be loaded. Now using the vertical scroll bar on the right-hand side of the panel move down on to the next wing, you will see in the **Navigation panel** whereabouts you are. When the first column of wings has been done move the horizontal scroll bar to approximately the middle to bring the next column into view and using the vertical scroll bar move up to the top of that column of wings on to the top wing of the column.

Repeat the procedure until all wings have been done in the sample. If one is using 15 wings in a file it will mean loading the next file with the rest of the wings and working through them as above.

Once all 30 wings are done just click on the **Graph icon** on the tool bar and Hey Presto! The results appear in a spreadsheet complete with graphs.

The panels appear on the monitor screen similar to the arrangement shown on the next page, this gives some idea of how the computer screen appears when using the Beemorph program.

This program is impressive in its ease of use and will appeal to beekeepers who are interested in using morphometry as a means of selection. It can be tried free of charge for 30 days, then by registering and paying the registration fee of £25 one can then continue in its use.

Russell Talbot is to be congratulated in designing such an excellent program.

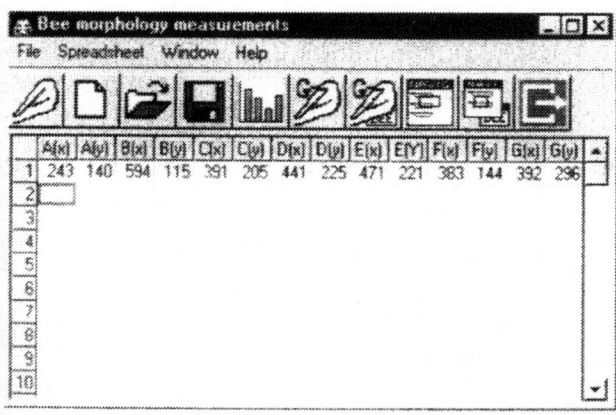

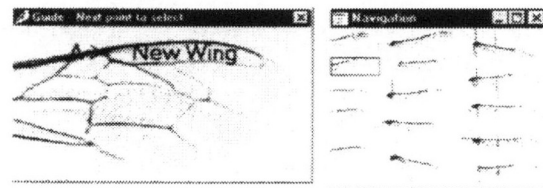

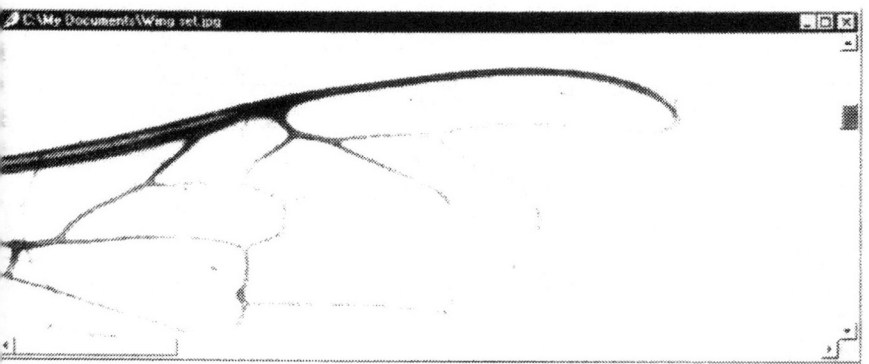

Fig 15   This shows how the monitor screen looks during operation.

Another computer program for use with morphometry is a Czechoslovakian program called 'Beewings', this is available from the Bee Research Institute at Dol, in Czechoslovakia. A demonstration copy can be obtained by sending an E-mail to Dalibor Titera at:

dalibor@beedol.cz

Alternatively a license copy costs 90 Euros.

Beewings morphometry programme
Instructions for use

Install the program from the CD to your hard disk. Then using the license disk get the license installed as per the instructions. If you only have a demonstration version there will be no license disk, and you will be limited to processing 5 wings at a time.

You can either use the wing files provided in the folder marked 'sample wings' or create your own wing files using a film scanner or a suitable flat bed scanner.

If you decide to use a scanner, scan your own wings before opening the Beewings programme. If you only want to measure 8 points on each wing to just get the D.S. and C.I. then instead of scanning the entire wing you can scan the part of the wing that contains the 8 points you will be using (see wing on page 60). This will give a larger image on the screen and make for greater accuracy. Skip the next section in italics if you are not going to use the scanner.

Scan the wings using a scanner with a minimum resolution of 1200 DPI, create a file for each of the wings by drawing out a frame that includes the whole wing, or if you only want to measure 8 points on the wing to produce cubital index and discoidal shift draw out a frame that is ' approximately 4mm X 7mm, (this will give a larger image of the wing on the monitor). Position the frame so that it is almost touching the front edge of the wing and has all the radial cell and the discoidal cell within it, and save it with a file name to a location you will remember, for example:-
My Documents/Mywings1
pull the frame down over the next wing and save it as:-
Mywings2 *and so on.*

Now open the Beewings programme by clicking on
Start/Programmes/Beewings.

Click File/create wing set. A panel will appear called **Input Dialogue**

Type in the name you are calling this set of wings, say, **Mywings** and **click O.K**. On the screen of **Wing Set Description** click on the number of points: **3,4,7,8, or All points**.

(If you only want to get Cubital Index and discoidal shift readings you select 7 or 8). Click on the **Identification tab** near the bottom of the panel and fill in the details of the colony, beekeeper etc if you want to record this information and have it printed out on the report.

Click on Properties and hit the **ADD** button and browse to find your wing images. **C: \My Documents**. (If you want to use the wing samples on the floppy disc put the floppy in A: drive and select A:\ ) The files will appear in the left-hand side of the browser box once you have located them.

Click on the first wing file hold down the shift key and highlight the whole set of wing files, release the shift key, Click O.K. the whole set will have gone through.

(If you do not have a license for the program you can only use 5 wings.)

Then click on **Save and lock set for making wings**. Get rid of the **Wing set Directories** panel by **clicking on the X** in the right hand corner of the panel.

Click on File/run wing set

Then a box will appear with:- start your session with this wing, **click OK**. Then **Click View/Marking Scheme**

Below your wing a map of a wing will appear showing the sequence of marking, you can position this map well clear of the wing by putting the cursor on the blue band at the top of the panel and hold down the mouse button and drag the panel to a new position. The map shows the points that have to be clicked on in the order they have to be done. It starts at 0 and goes up to 17 (if you selected All points) I should point out that the point marked 0 on the map was in the wrong place in the earlier version of Beewings, this was corrected later. It should be at the apex of the radial cell at the right-hand end of the radial cell. (See the wing displayed below)

Mark the wing in the sequence as indicated. Each time you mark a point a red spot will appear in the **Marking Scheme panel** showing you where the next point to be marked is located.

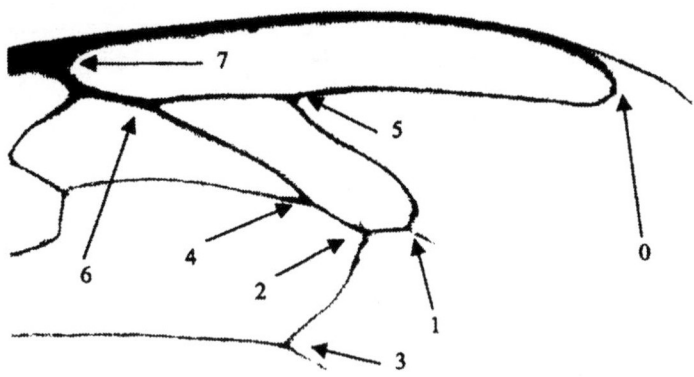

If you find you have made a mistake and your cross mark is not exactly in the correct place you have two options/clear all marks in **Wings/Clear all marks**, and start again from the paint marked 0, or click on the cross and it will turn red, move your cursor away from the place by a few inches and click again, the cross will turn blue, click again the; cross turns red and now put your cursor back in the correct place and click again, the red cross will disappear and your' new blue cross will be in the correct spot.

When all the points on a wing have been done

Click on wings/save marked wing

Click on the right-hand green arrow on the Tool bar and the next wing will appear.

Mark and repeat as above.

When the last wing has been marked, as you click on - **Save marked wing**, you will get the message, "you have completed, etc. and can generate results"

Click on the cross on the top right-hand of the screen to get rid of that panel and

Click on File/generate results.

Print out the page and click close, a second report page will appear if you selected All Points or 8 points, but if you selected 7 points only one report will be produced, click on print to print out the second report.

A better way of displaying the results however is to export the data produced by Beewings into the BIBBA spreadsheet that will display the data of cubital index and discoidal shift and create graphs from some of the data that is stored in
C: \Programs\beewings\wingsets\filename \csv

If you cannot see this file make sure you select All files in the panel near the bottom.

By clicking on the file you will find something like 104 columns of data if you chose the All Points option, otherwise many of the columns will, contain zeros.

You can, if you wish create graphs from this data.

One can for instance copy and paste the data for cubital index and discoidal shift into the BIBBA Morphometry spreadsheet and let it create the graphs automatically for cubital Index and discoidal shift. The cubital index appears in the Beewings data spreadsheet labeled CI in column CC the discoidal shift measurement in degrees is labeled Dis A in column CF. Highlight these columns separately by pressing **Control and C** and paste these into the BIBBA Morphometry sheet using **Control and V**.

The method of marking using the Beewings program appears to be more accurate than using the conventional method, and sitting at a computer is less tiring than measuring wings with a projector.

Using the scans at the magnification given by the frame size we recommend produces a large image of the part of the wing containing the points necessary to be measured. This also lends itself to measuring the indices by placing a transparent measuring device on the screen and physically measuring them as one would using a projector, and this is useful to compare the results with the data produced by

Beewings. Obviously this is something that need only to be done once if you want to convince yourself that the results you are getting from the Beewings program are correct.

A word of caution at this point, some beekeepers get carried away with this technology. They assume that because morphometry shows they have a pure-bred bee of a certain sub-species then this will make a good breeder colony. Such a colony although being pure-bred could have faults that could indicate it is not suitable for breeding. In other words assessments made during examinations are more important than the results using morphometry. Morphometry then should only be used on colonies that show up well in the records of colony assessment, this will then show which colonies that have the best assessment records are pure-bred ones. These can then be used as the queen mothers in the breeding process. It is important to understand that morphometry is not the 'magic bullet' but is really the last step in the selection process.

Other uses of computers in beekeeping.

Nowadays many beekeepers have computers and these can be of help in beekeeping. They can be used for colony records, queen rearing time-tables, producing graphs for plotting morphometry results.

BIBBA have several computer aids designed for some of the tasks mentioned, and these are available free to any beekeeper who sends a formatted floppy disk to me with a S.A.E., alternatively they can be down loaded from the BIBBA web site at:

http://www.bibba.com

Albert Knight
BIBBA Groups Secretary
11 Thomson Drive
Codnor
Ripley Derbyshire
DES 9RU

References

Cooper BA (1986)
Honeybees of the British Isles, 1986, BIBBA.

Dade HA (1962)
Anatomy and Dissection of the Honeybee, 1962, IBRA.

Möbus B (1983)
Mating in Miniature, 1983, BIBBA.
*Note that the recipe for queen cage candy here is incorrect.

Ruttner F (1988)
Breeding Techniques and Selection for Breeding of the Honeybee, 1988, BIBBA.

Ruttner F (1988a)
Biogeography and Taxonomy of the Honeybee, 1988, Springer-Verlag.

Ruttner F, Milner E and Dews JE (1990)
The Dark European Honey Bee, 1990, BIBBA.

Acknowledgements

We wish to express our thanks to many helpers, especially to Dr F Ruttner, Dr J van Praagh of Celle, Dr V Maul of Kirchhain and Dr F Schaper of Erlangen, for information on the simplified techniques and the interpretation of data employed in morphometry.

BIBBA – The British Isles Bee Breeders' Association

One very important aspect of BIBBA's work is in encouraging the formation and the work of breeding groups, who can do so much in identifying and propagating any strains of the native Dark bee in their area.

BIBBA supports such groups by training in morphology and in certain circumstances by the provision of genetic material.

Further information can be obtained either from the Membership Secretary:

> Mr BP Dennis
> Eastwood Apiary
> 50 StatioN Road
> Cogenhoe
> Northampton, NN7 1LU

or the General Secretary:

> Mr A Knight
> 11 Thomson Drive
> Codnor
> Ripley
> Derbyshire, DE5 9RU

BIBBA publishes books and booklets on various aspects of breeding, some of which are mentioned in this booklet. A comprehensive list can be obtained from the BIBBA Postal Sales Officer:

> Mr A.B. Hinchley
> BIBBA Postal Sales
> Meadow Croft Apiaries
> 2 Birchwood Road
> Alfreton
> Derbyshire DE55 7HB
> Tel. 01773 832086

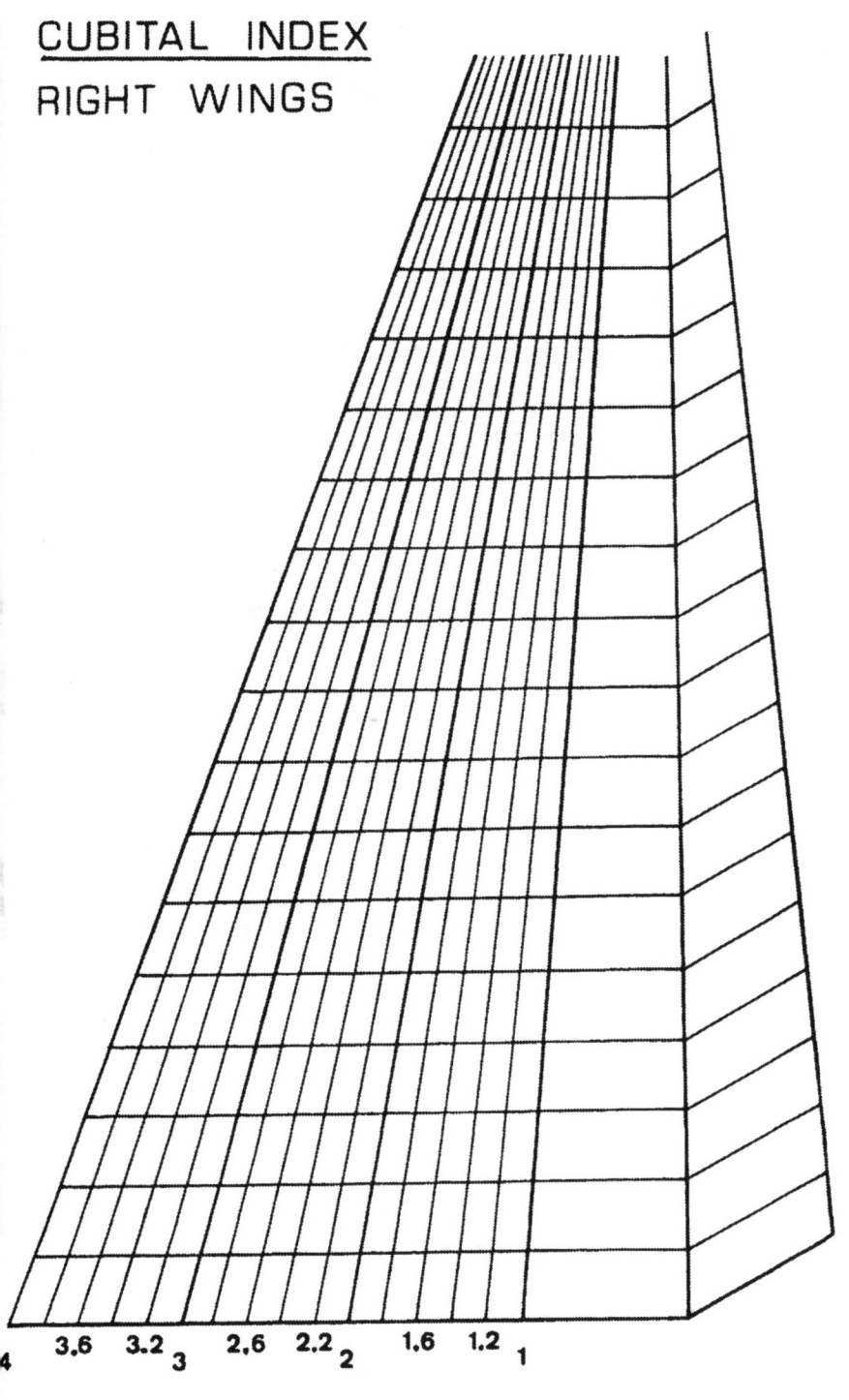

DISCOIDAL SHIFT

TOWARDS THE BODY
IS NEGATIVE

TOWARDS THE WING TIP
IS POSITIVE

4 2 0 2 4 6 8

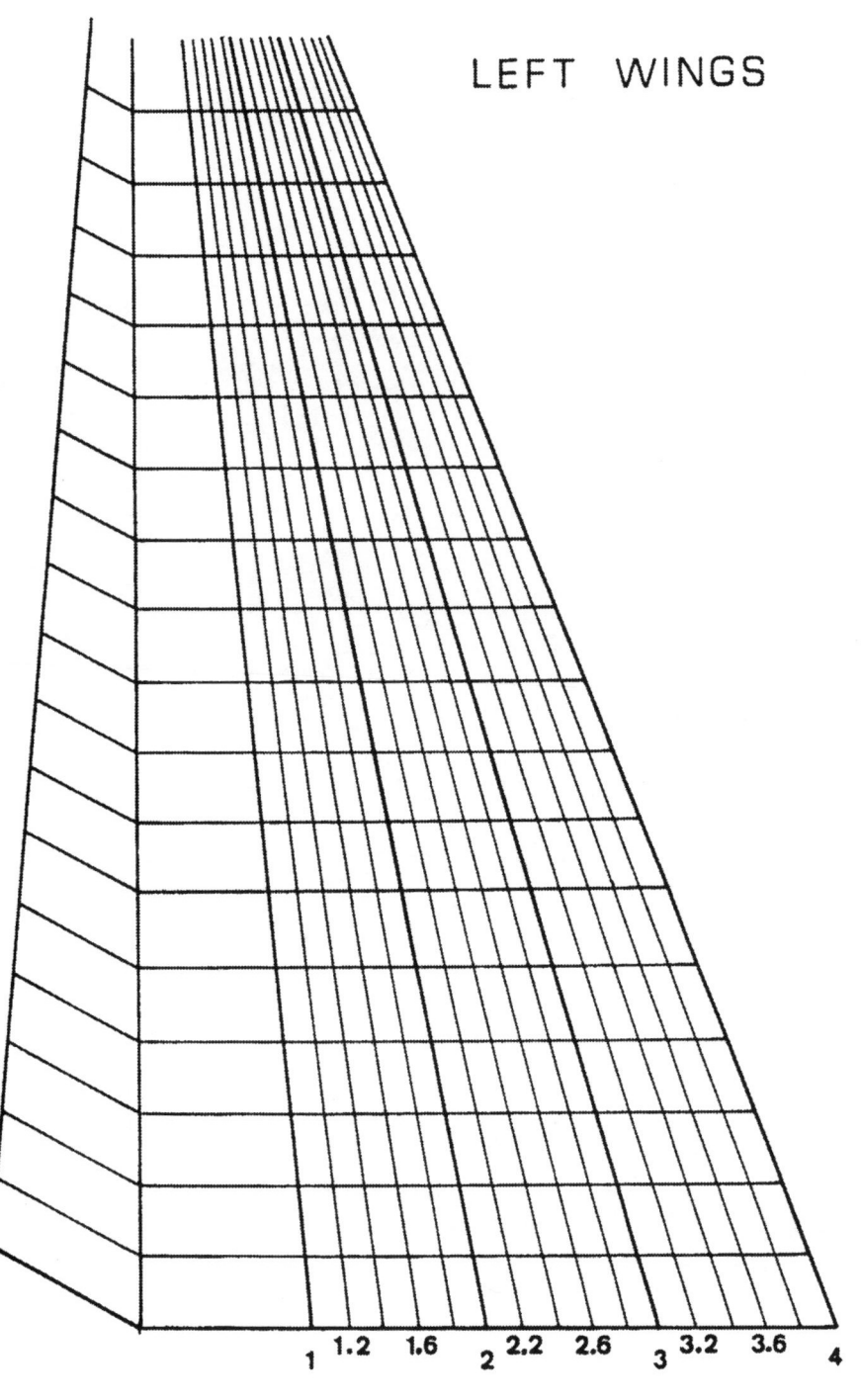